Studies in Automatic
Programming Logic

THE COMPUTER SCIENCE LIBRARY

Artificial Intelligence Series
NILS. J. NILSSON, *Editor*

Studies in Automatic Programming Logic

Zohar Manna
Richard Waldinger

with contributions by
Shmuel Katz and Karl Levitt

NORTH-HOLLAND·NEW YORK
NEW YORK • AMSTERDAM • OXFORD

Elsevier North-Holland, Inc.
52 Vanderbilt Avenue, New York, New York 10017

North-Holland Publishing Company
P.O. Box 211
Amsterdam, The Netherlands

Library of Congress Cataloging in Publication Data

Manna, Zohar.
   Studies in automatic programming logic.
   (Artificial intelligence series) (The
computer science library)
   Bibliography: p.
   Includes index.
   1. Electronic digital computers—Programming.
2. Computer programs. I. Waldinger, Richard,
joint author. II. Title.
QA76.6.M357   001.6'42   77-5765
ISBN 0-444-00224-3
ISBN 0-444-00225-1 pbk.

Contents

Preface

As computers have developed, they have come to play a larger and larger role in the programming process. Although twenty-five years ago programmers had to express their programs as sequences of numbers, and to interpret similarly encrypted error messages, today they are assisted by a battalion of assemblers, compilers, editors, interactive debuggers, and other tools by which the machine aids in its own programming. The next stage in the progression is likely to be the appearance of systems that can understand the subject matter of the program being constructed, and thus play a more critical and active role in its development. A compiler, for instance, can detect a syntactic error such as a missing *plus* sign, but it cannot hope to recognize a logical error, such as when a *plus* sign has been replaced by a *minus*. To detect such a mistake, a system must understand the program more thoroughly than the compiler does. In recent years, experimental systems with this level of understanding have begun to appear.

The three papers in this collection illustrate how intelligent systems can be applied to the verification, debugging, and synthesis of computer programs:

Reasoning About Programs (by R. Waldinger and K. Levitt) describes a computer system to prove the correctness of a given program.

Logical Analysis of Programs (by S. Katz and Z. Manna) presents techniques for the automatic documentation of given programs to assist in their debugging or verification.

Knowledge and Reasoning in Program Synthesis (by Z. Manna and R. Waldinger) suggests methods for the automatic development of computer programs.

This collection treats, a progression of tasks of increasing complexity. In the first paper, we attempt to show that a given program is consistent with given specifications and documentation, in the form of "invariant assertions." In the second paper, we begin with only the program and its specifications, and we introduce means to generate the documentation of the program and to correct the program if the specifications are not met. In the final paper, we assume that only the specifications are given, and we propose techniques for developing a program guaranteed to meet the specifications.

Progress in the implementation of these techniques reflects the increasing complexity of their corresponding tasks and the ingenuity they require. Program verification has been studied intensely, and our first paper actually describes a running verification system. (The description of the system is exceptionally complete, containing as an appendix a full annotated listing of the theorem prover and traces of sample solutions. Thus, a reader who is interested in actually implementing a verification system can see exactly how it was done.) On the other hand, debugging and synthesis are not yet so well understood, and the methods in our second and third papers have only been partially implemented.

The three papers have appeared earlier in journals, but are presented here in a corrected, updated, and slightly modified form. "Reasoning about programs" appeared in the journal *Artifical Intelligence* (Vol. 5, No. 3, Fall 1974, pp. 235–316), "Logical analysis of Programs" in the *Communications of the ACM* (Vol. 19, No. 4, Apr. 1976, pp. 188–206), and "Knowledge and reasoning in program synthesis" in *Artifical Intelligence* (Vol. 6, No. 2, Summer 1975, pp. 175–208). We are indebted to these journals for permission to reprint the papers in this collection. In the subsequent postscript we summarize more recent developments briefly. The references have been combined into a single bibliography at the end.

Z. M. and R. W.

Chapter 1
Reasoning About Programs

Richard Waldinger and Karl Levitt

Problems worthy
of attack
Prove their worth
by hitting back.
Piet Hein

1. Introduction and Background

This paper describes a computer system that proves theorems about programs, a task of practical importance because it helps certify that the programs are correct. Instead of testing a program on test cases, which may allow some bugs to remain, we can try to prove mathematically that it behaves as we expect.

Many programs have done this sort of reasoning. King's (1969) program verifier proved an interesting class of theorems and was very fast. Deutsch's (1973) system is perhaps not as fast as King's, but it can prove more interesting theorems. Igarashi et al. (1975) applied a resolution theorem prover to the verification of programs written in PASCAL, such as Hoare's (1961) FIND. Their system does little actual resolution but a lot of simplification and reasoning about equality. A program devised by Boyer and Moore (1975) can prove difficult theorems about LISP programs.

Thus there is no shortage of interesting work related to our own. The special characteristic of our system is that it is markedly

This is a revised version of a previously published article by the same name which appeared in *Artificial Intelligence,* vol 5, pp. 235-316. Copyright 1974 by North-Holland Publishing Company, Amsterdam. Reprinted here by permission of publisher.

concise, readable, and easy to change and apply to new subject areas.

Our program verifier consists of a *theorem prover* (or *deductive system)* and a *verification condition generator.* The verification condition generator takes an annotated program as input and constructs a list of theorems as output. The truth of the constructed theorems implies the correctness of the program. The task of the deductive system is to prove these theorems. The verification condition generator (Elspas et al., 1973) is written in INTERLISP (Teitelman, 1975), and the deductive system is written in QA4 (Rulifson et al, 1972). This paper focuses on the deductive system but, to be complete, gives examples of verification condition generation as well.

In writing our deductive system, we were motivated by several goals. First, the system should be able to find proofs; it should have enough deductive power to prove, within a comfortable time and space, the theorems being considered. Also, these proofs should be at the level of an informal demonstration in a mathematical textbook. This means that the difficulty in following one line to the next in any proof should be small enough that the proof is understandable, yet large enough not to be trivial. In any practical program verifier, the user will wish to follow the steps in a deduction. Furthermore, the strategies the system uses in searching for a proof should be strategies that we find natural. Not only should the tactics that eventually lead to the proof be ones we might use in proving the statement by hand, but also the false starts the system makes should be ones we might make ourselves. We do not want the system to rely on blind search; the trace of an attempted solution should make interesting reading.

In addition to the requirement that proofs be readable, the rules the system uses in going from one line to the next should be easy to read and understand. We should be able to look at a rule and see what it does. Also, it should be easy to change old rules and to add new rules. The user of a program verifier is likely to introduce new concepts, such as operators or data structures. We want to be able to tell the deductive system how these structures behave and to have the system reason effectively using the new symbols. Giving the system new information should be possible without knowing how the system works, and certainly without reprogramming it. Furthermore, the addition of new information should not degrade the performance of the system prohibitively.

The system is intended to evolve with use. As we apply the system to new problems, we are forced to give it new information and, perhaps, to generalize some old information. These changes are incorporated into the system, which may then be better able to solve new problems.

Since the system is easy to extend and generalize, we do not worry about the completeness or generality of any particular version of the system. It is powerful enough to solve the sort of problem on which it has been trained, and it can be easily changed when necessary.

These considerations played a part in the design of the programming system called QA4, as well as the construction of our deductive system, which is written in the QA4 language. Some of the techniques described below are embedded in the QA4 system itself; others are expressed as parts of the deductive system.

2. The Method

Perhaps not all readers are familiar with the method of proving statements about programs that we have followed in our work. Our method is a natural technique introduced independently by Floyd (1967) and Naur (1966) and formalized by Hoare (1969). Knuth (1968, pp 17–18) traces the germ of the idea back to von Neumann and Goldstine (1963). An other presentation of the same idea appears in a lecture by Turing (1950). Although we cannot give a thorough introduction to the technique here, we provide below an example of its application to convey the flavor of the approach.

2.1. A Straight-Line Program

Consider a simple program, written in a flowchart language, that exchanges the values of two variables, as shown in Figure 1.1. We assume that before the program is executed, X and Y have some initial values X_0 and Y_0. Suppose we want to prove that after the program is executed, $X = Y_0$ and $Y = X_0$. Then we offer the *input* and *output assertions* as comments in our program shown in Figure

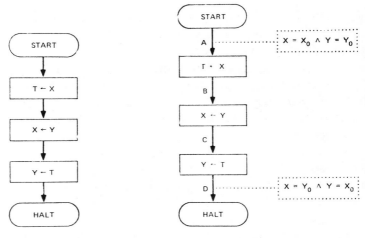

Figure 1.1.   Figure 1.2.

1.2. These assertions are not to be executed by the program, but they tell us something about the way the programmer expects his program to behave. He expects the assertion at A to be true when control passes through A, and the assertion at D to be true when control passes through D.

The essence of the approach is to generate from a commented program like the one above a set of statements called the *verification conditions*. If these statements and the input assertion are true, then the other assertions the programmer has put in his program are correct. Whereas the programmer's assertions are intended to hold only when control passes through the appropriate point, the verification conditions should be true in general, and they can be proved by a deductive system that knows nothing about sequential processing, loops, recursion, or other concepts about the flow of control of the particular program.

To generate the verification condition for our sample program, we pass the output assertion back toward the input assertion. As we pass it back, we change it to reflect the changing state of the system. In particular, if any assignments are made within the program, then the corresponding substitution should be made in the assertion. Passing the assignment at D back to point C changes it to $X = Y_0$ and $T = X_0$, as shown in Figure 1.3. We can argue that if the assertion at C

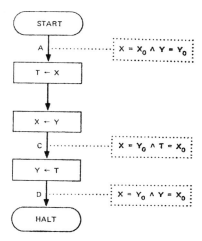

Figure 1.3.

is true when control passes through C, then the assertion at D will be true when control passes through D. In particular, if $T = X_0$ is true at C, and we execute $Y \leftarrow T$, then $Y = X_0$ will be true at D.

Passing the assertion all the way back to A in this manner gives the assertion $Y = Y_0 \wedge X = X_0$. If this assertion is true at A, then the final assertion will be true at D. However, we are already given the initial assertion $X = X_0 \wedge Y = Y_0$. The truth of the assertion at D then depends on the truth of the obvious implication $X = X_0 \wedge Y = Y_0 \supset Y = Y_0 \wedge X = X_0$. This statement is the verification condition for this program. It can be proved by a deductive system independently of any knowledge about this program.

Constructing verification conditions by this method is an algorithmic process, not a heuristic one. On the other hand, there can be no cut and dried algorithm for proving such verification conditions. However, the somewhat restricted domain of program verification is more tractable than the general theorem-proving problem.

2.2. A Loop Program

Before we explain how the system is structured or implemented, let us first look at some examples of how verification conditions are generated and proved by our system. This example will give a better idea of the subject domain of the inference system and of the sort of reasoning it has to do.

Suppose we are given the annotated program shown in Figure 1.4 to compute the largest element in an array and its location. This program searches through the array, keeping track of the largest element it has seen so far and the location of this element. The intermediate assertion[1] at C says that MAX is the largest element in the array between 0 and I and that LOC is the index for MAX. Although our assertion language does not permit the ellipsis notation ("..."), we have introduced some suitable analogues, which are discussed later.

To prove assertions about a complex program, the system decomposes it into simple paths. This program can be decomposed into four simple paths:

(1) The path from B to C.
(2) The path from C to D.
(3) The path from C around the loop and back to C through point E.
(4) The path from C around the loop and back to C through point F.

Notice that the author of this program has put assertions not only at the START and HALT nodes of the program, but also at the intermediate point C. He has done this so that the proof of the program can be reduced to proving straight-line paths in the same way that the simple program of the previous section was verified. For instance, the path that begins at C, travels around the loop through E, and returns to C can be regarded as a simple, straight-line program with the assertion at C as both its start assertion and its halt assertion. The assertion at C has been cleverly chosen to be true when the loop is entered and to remain true whenever control travels around the loop and returns to C, and to allow the assertion at D to be proved when control leaves the loop and the program halts. The choice of suitable internal assertions can be an intellectually exacting

---

[1] In this program, and in examples throughout the paper, when we list several statements in an assertion, we mean the implicit conjunction of those statements. We will often also refer to each conjunct as an "assertion".

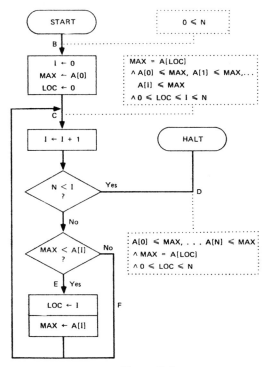

Figure 1.4.

task; some heuristic methods have been proposed that will work in this and many other examples (Elspas, 1974; Wegbreit, 1974; German and Wegbreit, 1975; Katz and Manna, 1976).

If all the straight-line paths of the program are shown to be correctly described by the given assertions, and if the program can be shown to terminate (this must be done separately), then we can conclude that the program is indeed correct, at least with respect to the programmer's final assertion.

Although there are many paths in the decomposition of a program, typically most of the paths are easy to verify. For this program, we examine two of the paths.

2.2.1. Verifying One Path

First, suppose we want to demonstrate that if the assertion at point C is true when control passes through C, then the assertion at

C will still be true if control passes around the loop and returns again to C. We will restrict our attention to the more interesting case, in which the test MAX < A[I]? is true; in this case, control passes through E. Furthermore, we will try to prove only that the second conjunct of the assertion at C remains true. Our verification condition generator gives us the following statement to prove:

$$MAX = A[LOC] \quad \wedge \tag{1.1}$$

$$A[0] \leqslant MAX, ..., A[I] \leqslant MAX \quad \wedge \tag{1.2}$$

$$0 \leqslant LOC \leqslant I \leqslant N \quad \wedge \tag{1.3}$$

$$\neg (N < I + 1) \quad \wedge \tag{1.4}$$

$$MAX < A[I + 1] \quad \supset \tag{1.5}$$

$$A[0] \leqslant A[I + 1], ..., A[I + 1] \leqslant A[I + 1]. \tag{1.6}$$

This statement is actually represented as five separate hypotheses and a goal to be deduced from these hypotheses. Lines (1.1) through (1.3) come from the assertion at C, and lines (1.4) and (1.5) come from the tests along the path. Line (1.6) comes from the assertion at C again. How the above statement is derived from the program is shown in detail below.

The behavior of the deductive system in this problem is typical of its approach to many problems. The goal (1.6) is broken into two subgoals:

$$A[0] \leqslant A[I + 1] \quad \wedge \quad \cdots \quad \wedge \quad A[I] \leqslant A[I + 1] \tag{1.7}$$

and

$$A[I + 1] \leqslant A[I + 1]. \tag{1.8}$$

The second subgoal, (1.8), is immediately seen to be true. The first subgoal, (1.7), is easily derived from (1.2) and (1.5).

2.2.2. Generating a Verification Condition

For those readers unfamiliar with the Floyd method of producing verification conditions, we give an example of its application: a complete trace of how the above verification condition was produced.

The path under consideration begins at point C, travels around the loop through point E, and returns again to C. We will try to prove the second conjunct at C.

This statement is

$$A[0] \leqslant MAX, \quad A[1] \leqslant MAX, \quad ..., \quad A[I] \leqslant MAX. \quad (1.9)$$

We pass this assertion backward around the loop to point E, making the corresponding substitution. The transformed assertion is then

$$A[0] \leqslant A[I], \quad A[1] \leqslant A[I], \quad ..., \quad A[I] \leqslant A[I]. \quad (1.10)$$

Since LOC does not appear explicitly in (1.10), the assignment LOC ← I has no effect.

To reach point E, the test

$$MAX < A[I]? \quad (1.11)$$

must have been true. Passing the assertion back before the test gives the implication

$$MAX < A[I] \quad \supset \quad A[0] \leqslant A[I], A[1] \leqslant A[I], ..., A[I] \leqslant A[I]. \quad (1.12)$$

If this implication is true before the test (1.11), the assertion (1.10) will be true after the test. To travel around the loop at all, the result of the test

$$N < I? \quad (1.13)$$

must have been false. Passing the assertion (1.12) back over the test (1.13) gives

$$\neg (N < I) \wedge MAX < A[I]$$
$$\supset \quad A[0] \leqslant A[I], A[1] \leqslant A[I], ..., A[I] \leqslant A[I]. \quad (1.14)$$

Passing (1.14) back over the assignment statement

$$I \leftarrow I + 1 \quad (1.15)$$

gives

$$\neg (N < I + 1) \wedge MAX < A[I + 1]$$
$$\supset \quad A[0] \leqslant A[I + 1], A[1] \leqslant A[I + 1], ..., A[I + 1] \leqslant A[I + 1]. \quad (1.16)$$

This statement has been generated in such a way that if it is true when control passes through point C, then (1.9) will be true if control passes around the loop through point E and returns to C. If we consider this path as a straight-line program with the assertion at C as both its start assertion and its halt assertion, then proving the correctness of the second conjunct (1.9) at C reduces to proving

$$MAX = A[LOC] \quad \wedge$$

$$A[0] \leqslant MAX, ..., A[I] \leqslant MAX \quad \wedge$$

$$0 \leqslant LOC \leqslant I \leqslant N \quad \wedge$$

$$\neg (N < I + 1) \quad \wedge$$

$$MAX < A[I + 1] \quad \supset$$

$$A[0] \leqslant A[I + 1], ..., A[I + 1] \leqslant A[I + 1].$$

Finally, the antecedents of this implication are expressed as separate hypotheses, and the consequent is represented as a goal. This is exactly the condition that was proved in the previous section.

2.2.3. Verifying Another Path

Now let us look more briefly at the path from C to D. We will assume the assertion at C is true and will prove the assertion at D. We will look at the first conjunct of the assertion at D. Our verification condition generator gives us the following statement to prove:

$$MAX = A[LOC] \quad \wedge \qquad\qquad (1.17)$$

$$A[0] \leqslant MAX \wedge \cdots \wedge A[I] \leqslant MAX \quad \wedge \qquad (1.18)$$

$$0 \leqslant LOC \leqslant I \leqslant N \quad \wedge \qquad\qquad (1.19)$$

$$N < I + 1 \quad \supset \qquad\qquad (1.20)$$

$$A[0] \leqslant MAX \wedge \cdots \wedge A[N] \leqslant MAX. \qquad (1.21)$$

The reasoning required for this proof is a little more subtle than the previous deduction. When the system learns that $N < I+1$ (1.20), it immediately concludes that $N + 1 \leqslant I + 1$, since N and I are integers. It further deduces that $N \leqslant I$. Since it already knows that $I \leqslant N$ (1.19), it concludes that $N = I$. Using the hypothesis (1.18), the system reduces the goal (1.21) to proving that $I = N$, which it now knows.

This deduction involves a lot of reasoning forward from assumptions, while the preceding deduction required reasoning backward from goals. Both of these proofs are typical of the behavior of the system at large in their use of the properties of equality and the ordering relations.

In reading the QA4 listing of the theorem prover (Section 6), one is struck by the absence of any general deductive mechanisms outside of the language processor itself. The QA4 system incorporates enough of the common techniques of theorem proving and problem solving that our inference system needs no general problem-solving knowledge, but only some knowledge about numbers, arrays, and other structures. The following sections show how the QA4 language allows that knowledge to be represented.

3. The Language

3.1. Pattern Matching and the Goal Mechanism

The deductive system is made up of many rules expressed as small functions or programs. Each of these programs knows one fact and the use for that fact. The QA4 programming language is designed so that all these programs can be coordinated; when a problem is presented to the system, the functions that are relevant to the problem "stand forward" in the sense explained below.

A program has the form

$$(\text{LAMBDA } \langle \text{pattern} \rangle \langle \text{body} \rangle).$$

Part of the knowledge of what the program can be used for is expressed in the pattern. When a function is applied to an argument, the pattern is matched against that argument. If the argument turns out to be an instance of the pattern, the match is said to be successful. The unbound variables in the pattern are then bound to the appropriate subexpressions of the argument, and the body of the program is evaluated with respect to those new bindings.

For example, the program

$$\text{REVTUP} = (\text{LAMBDA } (\text{TUPLE } \leftarrow X \leftarrow Y) (\text{TUPLE } \$Y \$X))$$

has pattern (TUPLE $\leftarrow X \leftarrow Y$) and body (TUPLE $\$Y \$X$). The symbol "TUPLE" is treated as a constant, and can only be matched against itself. The prefix "$\leftarrow$" means that the variable is to be given a new binding. The prefix "$" means that the variable's old binding is to be used. When REVTUP is applied to (TUPLE A B), the pattern (TUPLE ←X ←Y) is matched against (TUPLE A B). The match is seen to be successful, the variable X is bound to A and the variable Y is bound to B. The body (TUPLE $Y $X) is evaluated with respect to these bindings, giving (TUPLE B A).

On the other hand, if a function is applied to an argument and the pattern of that function does not match the argument, a condition known as *failure* occurs. At many points in the execution of a program, the system makes an arbitrary choice between alternatives. Failure initiates a backing up to the most recent choice and the selection of another alternative, if one exists. The mismatching of patterns is only one of the ways in which failure can occur in a program.

We have yet to explain how a program stands forward when it is relevant. In the above example, the function was called by name, much as it is in a conventional programming language. But it is also possible to make an argument available to any applicable program in a specified class. This is done by means of the *goal mechanism*.

When we say (GOAL ⟨goalclass⟩ ⟨argument⟩), where the goal class is a tuple of names of functions, we first check to see whether the argument is known to be true, in which case the statement succeeds. Otherwise that argument becomes available to the entire goal class. The pattern of each of those functions is matched in turn against the argument. If the match is successful, the function is applied to that argument. If the function returns a value, that value is returned as the value of the goal statement. On the other hand, if the match of the pattern fails, or if a failure occurs in evaluating the function, backtracking takes place, the next function in the goal class is tried, and the process is repeated. If none of the functions in the goal class succeed, the entire goal statement fails.

For example, in our deductive system, one of the goal classes is called EQRULES, the rules used for proving equalities. One of these rules is

EQTIMESDIVIDE =
    (LAMBDA (EQ ←W (TIMES (DIVIDE ←X ←Y) ←Z))
    (GOAL $EQRULES
        (EQ (TIMES $Y $W)(TIMES $X $Z)))).

This rule states that to prove W = (X/Y)*Z, we should try to prove Y*W = X*Z. It is assumed (for simplicity) that Y will not be 0. (The actual EQTIMESDIVIDE, shown in Section 6, is more general than this.) The rule has the pattern

(EQ ←W (TIMES (DIVIDE ←X ←Y) ←Z)).

If we execute (GOAL $EQRULES (EQ A (TIMES (DIVIDE B C) D))) [i.e., we want to prove A = (B/C)*D], the system will try all the applicable EQRULES in turn. If none of the previous rules succeed, the system will eventually reach EQTIMESDIVIDE. It will find that the pattern of EQTIMESDIVIDE matches this argument, binding W to A, X to B, Y to C, and Z to D. Then it will evaluate the body of this function; i.e., it will try

(GOAL $EQRULES (EQ (TIMES A C) (TIMES B D))).

If it succeeds at proving (EQ (TIMES A C) (TIMES B D)), it will return normally. If it fails, it will try to apply the remaining EQRULES to the original argument, (EQ A (TIMES (DIVIDE B C) D)). The goal statement is an example of the pattern-directed function invocation introduced by Hewitt (1971) in PLANNER.

The net effect of this mechanism is that it enables the user to write his programs in terms of what he wants done, without needing to specify how he wants to do it. Furthermore, at any point, he can add new rules to EQRULES or any other goal class, thus increasing the power of the system with little effort.

3.2. Some Sample Rules

The deductive system is a collection of rules represented as small programs. One rule was given in the preceding section; two more rules are presented here. The complete deductive system is included in Section 6.

The first rule, EQSIMP, attempts to prove an equality by simplifying its arguments:

```
EQSIMP = (LAMBDA (EQ ←X ←Y)
          (PROG (DECLARE)
               (SETQ ←X ($SIMPONE $X))
               (GOAL $EQRULES (EQ $X $Y)))
          BACKTRACK).
```

This rule says: to prove an equality, try to simplify one side of the equality. The rule, a member of EQRULES, has the pattern (EQ ←X ←Y). It is applicable to a goal of form (EQ A B), where A and B are any expressions. When EQSIMP is applied to such a goal, X will bound to A and Y to B. SIMPONE, the simplifier, will simplify A. The goal statement tries to prove that the simplified A is equal to B. If it succeeds, the rule will return in the ordinary way. If the goal statement fails, or if the simplifier fails to simplify A, the entire application of the rule will fail.

The predicate EQ implicitly takes a set as its argument (see Section 3.4). Thus, there is an alternative match of (EQ ←X ←Y) to (EQ A B), binding X to B and Y to A. The user has specified the BACKTRACK option, meaning that he wants to try all possible matches. Therefore, if the first application of EQSIMP fails, the system will apply it the other way and try to simplify B. Only if this second attempt fails will the entire rule fail, allowing other members of EQRULES to work on the same goal.

The second rule is

FSUBTRACTI = (LAMBDA (←F (SUBTRACT ←X ←Y) ←Z)
                          (GOAL $INEQUALITIES
                              ($F $X (PLUS $Y $Z)))).

This rule says: to prove $X - Y \leqslant Z$, try to prove $X \leqslant Y + Z$. It belongs to the goal class INEQUALITIES and is thus used not only for the predicate LTQ ($\leqslant$), but also for LT ($<$), GT ($>$), and GTQ ($\geqslant$). The variable F is bound to the appropriate predicate symbol when the pattern is matched against the goal.

3.3. Demons

The goal mechanism is used for reasoning backward from a goal. However, sometimes we want to reason forward from a statement. For example, suppose that whenever an assertion of the form $X \geqslant Y$ is asserted, we want to assert $Y \leqslant X$ as well. We do this by a QA4 mechanism known as the *demon*.

A demon is imagined to be a spirit that inhabits a hiding place, waiting until some specified event occurs, at which time it appears, performs some action, and vanishes again. We have put several demons in the system, each watching for a different condition. For instance, one demon watches for statements of the form $X \geqslant Y$ and makes the statement $Y \leqslant X$. The user of the system can create his own demons. Demons are a tool for reasoning forward from an antecedent. In particular, we use demons to drive antecedents into a canonical form. For example, we drive all inequality expressions with integer arguments into an assertion of the form $X \leqslant Y$.

3.4. Representations

To as great an extent as possible, we have chosen representations that model the semantics of the concepts we use so as to make our deductions shorter and easier. For example, our language has data structures especially intended to eliminate the need for certain inferences. In addition to *tuples,* which are like the familiar lists of the list-processing languages, we have the finite *sets* of conventional mathematics, and *bags,* which are unordered tuples or, equivalently, sets that may have multiple occurrences of the same element. [Bags are called multisets by Knuth (1969), who outlines many of their properties.] Furthermore, we allow arbitrary expressions to have property lists in the same way that atoms can have property lists in LISP (McCarthy et al., 1962).

These data structures are useful in the modeling of equivalence relations, ordering relations, and arithmetic functions. For instance, if the addition of numbers and the multiplication of numbers are each represented by a function of two arguments, then it becomes necessary to use numerous applications of the commutative and associative laws to prove anything about the number system. However, in QA4 all functions take only one argument, but this argument can be a tuple, set, or bag, as well as any other expression. Functions of multiple arguments can be represented by a function defined on tuples. However, a function that is commutative and associative, such as PLUS, is defined on bags. The expression (PLUS A 2 B) really means (PLUS (BAG A 2 B)). Recall that bags are unordered; the system cannot distinguish between (BAG A 2 B) and (BAG 2 A B). Consequently, the expressions (PLUS A 2 B) and (PLUS 2 A B) are identically equal in our system. This makes the commutative law for addition redundant and, in fact, inexpressible in the language. Most needs for the associative law are also avoided.

The logical function AND has the property that, for instance, (AND A A B) = (AND A B). The number of occurrences of an argument does not affect its value. Consequently, AND takes a set as its argument. Since (SET A A B) and (SET A B) are indistinguishable, (AND A A B) and (AND A B) are identical, and a statement of their equality is unnecessary. Some functions that take sets as arguments are AND, OR, EQ, and GCD (greatest common divisor).

When a new fact is asserted to our system, the value TRUE is placed on the property list of that fact. If at some later time we want to know if that fact is true, we simply look on its property list.

However, certain facts are given special handling in addition. For example, if we tell the system that certain expressions are equal, we form a set of those expressions. On the property list of each expression, we place a pointer to that set. For instance, if we assert (EQ A B C), the system stores the following:

If we subsequently discover any of these expressions to be equal to still another expression, the system adds the new expression to the previously formed set and puts the set on the property list of the new expression as well. For instance, if we assert (EQ B D), our structure is changed to the following:

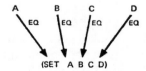

The transitivity, symmetry, and reflexivity of equality are thus implicit in our representation. If we ask whether A and D are equal, the system knows immediately by looking at the property list of A or D.

Ordering relations are also stored using the property-list mechanism. If we know that some expression A is less than B, we place a pointer to B on the property list of A:

$$A \xrightarrow{\text{LT}} B.$$

If we learn that B is less than C, we put a pointer to C on the property list of B:

$$A \xrightarrow{\text{LT}} B \xrightarrow{\text{LT}} C.$$

If we then ask the system if A is less than C, it will search along the pointers in the appropriate way to answer affirmatively. The transitive law is built into this representation.

The system knows about LT ($<$), GT ($>$), LTQ ($\leqslant$), GTQ ($\geqslant$), EQ (=), NEQ ($\neq$), and how these relations interact. For example, if we assert $X \geqslant Y$, $Y \geqslant Z$, and $X \leqslant Z$, the system will know $X = Y = Z$ and that (F X A) = (F Y A) for any function symbol F and argument A. Or if we assert $X \geqslant Y$ and $X \neq Y$, the system will know $X > Y$.

3.5. Contexts

When we are trying to prove an implication of the form A $\supset$ B, it is natural to want to prove B under the hypothesis that A is true. Our assumption of the truth of A holds only as long as we are trying to prove B; after the proof of B is complete, we want to forget that we have assumed A. For this and other reasons, the QA4 language contains a context mechanism. All assertions are made with respect to a context, either implicitly or explicitly. For any context, we can create an arbitrary number of lower contexts.

A query made with respect to a context will have access to all assertions made with respect to higher contexts but not to any assertions made with respect to any other contexts. For instance, suppose we are trying to prove $i < j \supset i + 1 \leqslant j$ with respect to some context C_0. We may have already made some assertions in context C_0. We establish a lower context, C_1, and assert $i < j$ with respect to C_1. Then we try to prove $i + 1 \leqslant j$ with respect to C_1. When proving $i + 1 \leqslant j$, we know $i < j$, as well as all the assertions we knew previously in C_0. When the proof of $i + 1 \leqslant j$ is complete, we may have other statements to prove in C_0. In doing these proofs, we will know all the assertions in C_0 and also, perhaps, the assertion $i < j \supset i + 1 \leqslant j$, but we will not know $i < j$ because it was asserted with respect to a lower context.

3.6. User Interaction

Sometimes our rules ask whether they should continue or fail. This allows a user to cut off lines of reasoning that he knows in advance are fruitless. If he makes a mistake in aswering the question, he may cause the system to fail when it could have succeeded. However, he can never cause the system to find a false or erroneous proof.

In addition to these mechanisms, which are built into the language processor, we have developed some notations that make it easier to describe programming constructs; these notations are a part of our assertion language and are interpreted by the deductive system.

3.7. Notation

In speaking about the program to find the maximum element of an array, we found it convenient to use the ellipsis notation ("..."). We have not introduced this notation into our assertion language; however, we have found ways of getting around its absence.

3.7.1. TUPA, SETA, BAGA

Let A be a one-dimensional array and I and J be integers. Then (TUPA A I J) is the tuple

$$(\text{TUPLE A}[I], A[I + 1], ..., A[J]).$$

If $I > J$, then (TUPA A I J) is the empty tuple.

(SETA A I J) and (BAGA A I J) are the corresponding bag and set. To state that an array is sorted between 0 and N, we assert

$$(\text{LTQ (TUPA A 0 N)}).$$

To state that an array A is the same in contents between 0 and N as the initial array A_0, although these contents may have been permuted, we assert

$$(\text{EQ (BAGA A 0 N) (BAGA } A_0 \text{ 0 N)}).$$

3.7.2. The STRIP Operator

Let X be a set or bag, $X = (SET\ X_1, ..., X_n)$, or $X = (BAG\ X_1, ..., X_n)$. Then (LTQ (STRIP X) Y) means $X_1 \leqslant Y$ and ... and $X_n \leqslant Y$. For instance, to state that MAX is greater than or equal to any element in an array A between I and J, we assert

$$(LTQ\ (STRIP\ (BAG\ A\ I\ J))\ MAX).$$

This is perhaps not quite so clear as

$$A[I] \leqslant MAX,\quad A[I+1] \leqslant MAX,\quad ...,\quad A[J] \leqslant MAX,$$

but we prefer it to the first-order predicate calculus notation,

$$(\forall u)[(I \leqslant u \wedge u \leqslant J)\quad \supset\quad A[u] \leqslant MAX].$$

The STRIP operator is also used to remove parentheses from expressions:

$$(BAG\ A\ (STRIP\ (BAG\ B\ C\ D)))$$

is

$$(BAG\ A\ B\ C\ D).$$

We will eventually need two distinct operators, one to act as a quantifier and one to remove parentheses, but the single operator STRIP has played both roles so far.

3.7.3. Access and Change

Arrays cannot be treated as functions, because their contents can be changed, whereas functions do not change their definitions. Thus, while $f(x)$ is likely to mean the same thing for the same value of x at different times, $A[x]$ is not. We overcome this difficulty by adopting McCarthy (1962) functions[2] ACCESS and CHANGE in our

---

[2] They are actually called c and a ("contents" and "assign") in that paper. We follow King (1969), who inadvertently reversed the roles of the initials.

explication of the array concept:

> (ACCESS A I) means A[I].
> (CHANGE A I T) means the array A after the assignment
>   statement A[I]←T has been executed.

We do not propose that ACCESS and CHANGE be used in writing programs or assertions; we do find that they make reasoning about arrays simpler, as King suspected they would.

The next section shows examples of some fairly difficult proofs produced by the deductive system. The actual traces for some of these are included in Section 7.

4. Examples

4.1. The Real-Number Quotient Algorithm

Very little work has been done to prove properties of programs that work on the real numbers or the floating-point numbers, although there is no reason to believe such proofs could not be done. Figure 1.5 shows, for instance, a program (Wensley, 1958) to compute an approximate quotient Y of real numbers P and Q, where $0 \leqslant P < Q$. This is an interesting computationally plausible algorithm. It uses only addition, subtraction, and division by 2, and it computes a new significant bit of the quotient with each iteration.

The algorithm can be understood in the following way. At the beginning of each iteration, P/Q belongs to the half-open interval [Y, Y + D). It is determined whether P/Q belongs to the left half or the right half of the interval. if P/Q is in the right half, Y is reset to Y + D/2; otherwise, Y is let alone. Then D is halved. Thus P/Q remains in the interval [Y, Y + D), and Y becomes a better and better approximation for P/Q. Initially, Y is 0 and D is 1. The variables A and B retain certain intermediate values to make the computation more efficient.

We will consider here only one path through this program, i.e., the path around the loop that follows the right branch of the test P < A + B. We will prove only one loop assertion: $P < Y*Q + D*Q$. Our verification condition generator supplies us with the following

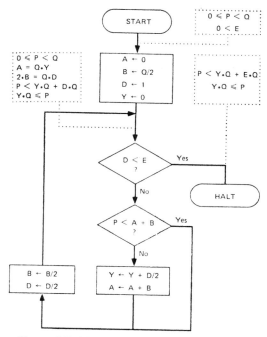

Figure 1.5. The Wensley Quotient Algorithm.

hypotheses:

$$0 \leqslant P, \tag{1.22}$$

$$P < Q, \tag{1.23}$$

$$A = Q*Y, \tag{1.24}$$

$$2*B = Q*D, \tag{1.25}$$

$$P < Y*Q + D*Q, \tag{1.26}$$

$$Y*Q \leqslant P, \tag{1.27}$$

$$\neg (D < E), \tag{1.28}$$

$$P < A + B. \tag{1.29}$$

The goal is to prove from these hypotheses that

$$P < Q*Y + Q*(D/2). \tag{1.30}$$

These hypotheses and the goal were constructed in a manner precisely analogous to the generation of the condition for the previous example of computing the maximum of an array.

The proof goes as follows. After an abortive attempt at using the assertion (1.26), the system tries to show that the conclusion follows from (1.29). It therefore tries to show that

$$A + B \leqslant Q*Y + Q*(D/2). \tag{1.31}$$

This goal is broken into the following two:

$$A \leqslant Q*Y, \tag{1.32}$$

$$B \leqslant Q*(D/2). \tag{1.33}$$

Of course, this strategy will not always be successful. However, in this case the goal (1.32) follows from (1.24), whereas (1.33) follows from (1.25).

A complete trace of this proof and listings of the rules required to achieve it are provided in the Appendices (Sections 6 and 7).

4.2. A Pattern Matcher

As an experiment in the incorporation of new knowledge into the system, we performed the partial verification of two new examples, a simple pattern matcher and a recursive version of the unification algorithm (Robinson, 1965). These algorithms were of special interest to us because they involve concepts similar to those actually used in the implementation of the QA4 system itself. They are thus in some sense realistic, although neither of these programs appears literally in the QA4 code.

Before we began proving properties of the pattern matcher we had only verified numeric algorithms. With the pattern matcher the system had to be acquainted with a new domain; it had to learn about expressions, substitutions, variables and constants. Therefore this phase of experiment tested the ability of the system to work with a new set of concepts. We will now describe this new domain.

We assume that *expressions* are LISP S-expressions (McCarthy et al., 1962); for example, (F X (G A B)) is an expression. Atomic elements are designated as either *constant* or *variable,* and they can be distinguished by the use of the predicates *const* and *var.* Here we use A, B, C, F, and G as constants and U, V, W, X, Y, and Z as variables:

var(X) is true,
const(A) is true,
var(A) is false,
var((X Y)) is false.

The predicate *constexp* is true if its argument contains no variables:

constexp((A B (X) C)) is false,
constexp((A B (D) C)) is true,
constexp(A) is true.

Note that *constexp*, as distinguished from *const*, can be true even if its argument is nonatomic.

A *substitution* replaces some of the variables of an expression by terms. Substitutions are represented as lists of dotted pairs. ((X · A) (Y · (F G))) is a substitution. *varsubst*(s, e) is the result of making substitution s in expression e. If s is

$$((X \cdot A) (Y \cdot (G B))),$$

and e is

$$(F X A (Y B)),$$

then *varsubst*(s, e) is

$$(F A A ((G B) B)).$$

The LISP functions *car, cdr, list,* and *atom* can be used to manipulate expressions. The empty substitution is denoted by EMPTY and has no effect on an expression. An operation called *compose,* the composition of substitutions, defined by Robinson (1965), has the following property:

varsubst(*compose*(s1, s2), e) = *varsubst*(s1, *varsubst*(s2, e)).

The problem of pattern matching is defined as follows: Given two expressions called the pattern and the argument, try to find a substitution for the variables of the pattern that makes it identical to the argument. We call such a substitution a *match.* For example, if the pattern *pat* is

$$(X\ (Y\ A\ B)\ X),$$

and the argument *arg* is

$$(D\ (C\ A\ B)\ D),$$

then *match(pat, arg)* is

$$((X \cdot D)\ (Y \cdot C)).$$

If there is no substitution that makes the pattern identical to the argument, we want the pattern-matcher to return the distinguished atom NOMATCH. Thus, if *pat* is (X Y X) and *arg* is (A B C), then *match(pat, arg)* = NOMATCH, since we cannot expect X to be matched against both A and C.

For simplicity, we assume that the argument contains no variables. A LISP-like program to perform the match might be

```
match(pat, arg) = prog((m1 m2)
     if const(pat) then (if pat = arg then return(EMPTY)
                                      else return(NOMATCH))
     if var(pat) then return(list(cons(pat, arg)))
     if atom(arg) then return(NOMATCH)
     m1←match(car(pat), car(arg))
     if m1 = NOMATCH then return(NOMATCH)
     m2←match(varsubst(m1, cdr(pat)), cdr(arg))
     if m2 = NOMATCH then return(NOMATCH)
     return(compose(m2, m1))).
```

The program does the appropriate thing in the case of atomic patterns or arguments, and it calls itself recursively on the left and right halves of the expressions in the nonatomic case. The program applies the substitution found in matching the left halves of the expressions to the right half of the pattern before it is matched, so as to avoid having the same variable matched against different terms.

We have proved several facts about a version of this program, but we focus our attention here on one of them: If the program does not return NOMATCH, then the substitution it finds actually is a match; i.e. applying the substition to the pattern makes that pattern identical to the argument. Thus, the output assertion is:

$$match(pat, arg) \neq \text{NOMATCH} \supset$$
$$varsubst(match(pat, arg), pat) = arg.$$

Since we assume the argument contains no variables, the input assertion is

$$constexp(arg). \tag{1.34}$$

We have verified one condition for the longest path of *match* with respect to these assertions. This path is followed when the pattern and the argument are both nonatomic and when the recursive calls on *match* successfully return a substitution. In writing our verification condition, we use the same abbreviations the program does, i.e.,

$$m1 = match(car(pat), car(arg)) \tag{1.35}$$

and

$$m2 = match(varsubst(m1, cdr(pat)), cdr(arg)).$$

In proving a property of a recursively defined program, we follow Manna and Pnueli (1970) and assume that property about the recursive call to the program. Thus, for this program we have the inductive hypotheses

$$constexp(car(arg)) \wedge m1 \neq \text{NOMATCH} \supset$$
$$varsubst(m1, car(pat)) = car(arg)$$

(the program works for the *car* of the pattern) and

$$constexp(cdr(arg)) \wedge m2 \neq \text{NOMATCH} \supset$$
$$varsubst(m2, varsubst(m1, cdr(pat))) = cdr(arg) \tag{1.36}$$

(the program works for the instantiated *cdr* of the pattern).[3] The verification condition generator would split both of these hypotheses into three cases; we will consider only the case in which the antecedents of both implications are true. Hence, we assume that both the recursive calls to the pattern matcher succeed in finding matches.

---

[3] Actually, in order to assume the inductive hypotheses we must ensure that the input assertion is satisfied for the recursive calls, in other words, that *constexp(car(arg))* and *constexp(cdr(arg))*. This follows immediately from the input assertion for the entire program, *constexp(arg)*.

By the path we have taken through the program, we know that

$$\neg const(pat) \qquad (1.37)$$

(the pattern is not a constant),

$$\neg var(pat) \qquad (1.38)$$

(the pattern is not a variable), and

$$\neg atom(arg) \qquad (1.39)$$

(the argument is not an atom). Since for this path

$$match(pat, arg) = compose(m2, m1),$$

the goal is to prove

$$varsubst(compose(m2, m1), pat) = arg. \qquad (1.40)$$

The proof produced by the system proceeds as follows. The goal is split into two subgoals:

$$varsubst(compose(m2, m1), car(pat)) = car(arg) \qquad (1.41)$$

and

$$varsubst(compose(m2, m1), cdr(pat)) = cdr(arg). \qquad (1.42)$$

From the property of *compose,* the first goal is simplified to

$$varsubst(m2, varsubst(m1, car(pat))) = car(arg).$$

Since

$$varsubst(m1, car(pat)) = car(arg)$$

by the first induction hypothesis (1.36), this simplifies to

$$varsubst(m2, car(arg)) = car(arg).$$

Since *arg* contains no variables, neither does *car(arg).* Thus, the goal simplifies to

$$car(arg) = car(arg).$$

The proof of (1.42) is even simpler:

$$varsubst(compose(m2, m1), cdr(pat))$$

simplifies to

$$varsubst(m2, varsubst(m1, cdr(pat))).$$

We know by our second induction hypothesis (1.36) that

$$varsubst(m2, varsubst(m1, cdr(pat))) = cdr(arg),$$

and this completes the proof.

This proof required not only that we add new rules describing the concepts involved, but also that we extend certain of our older capabilities, particularly our ability to simplify expressions using known equalities. A trace of the system's search for this proof is included in Section 7.

We worked nearly a week before the system was able to do this proof. However, once the proof was completed, the effort necessary to enable the system to do the proof of the unification algorithm was minimal. The latter proof, though longer than this one, did not require much additional intellectual capacity on the part of the deductive system. We do not show that proof here because it is similar to the pattern matcher proof, but we include the program and the assertion we proved about it.

4.3. The Unification Algorithm

The problem of unification is similar to that of pattern matching except that we allow both arguments to contain variables. We expect the algorithm to find a substitution that makes the two arguments identical when it is applied to both, if such a substitution exists. For example, if x is (F U A) and y is (F B V), then $unify(x, y)$ is $((U \cdot B) (V \cdot A))$, where U and V are variables and A, B, and F are constants.

A simple program to unify x and y is

```
unify(x, y) = prog((m1 m2)
    if x = y then return(EMPTY)
    if var(x) then
        return(if occursin(x, y) then NOMATCH
                                 else list(cons(x, y)))
    if var(y) then
        return(if occursin(y, x) then NOMATCH
                                 else list(cons(y, x)))
    if atom(x) then return(NOMATCH)
    if atom(y) then return(NOMATCH)
    m1 ← unify(car(x), car(y))
```

```
if m1 = NOMATCH then return (NOMATCH)
m2 ← unify(varsubst(m1, cdr(x)),
               varsubst(m1, cdr(y)))
if m2 = NOMATCH then return(NOMATCH)
return(compose(m2, m1))).
```

The predicate $occursin(u,v)$ tests if u occurs in v. This program is a recursive, list-oriented version of Robinson's iterative, string-oriented program. Again, we have verified only the longest path of the program, not the entire program. Furthermore, we have proved not the strongest possible statement about this program, but only that

$$unify(x,y) \neq \text{NOMATCH} \supset$$
$$varsubst(unify(x, y), x) = varsubst(unify(x,y), y).$$

4.4. The FIND Program

The program FIND, described by Hoare (1961), is intended to rearrange an array A so that all the elements to the left of a certain index F are less than or equal to A[F], and all those to the right of F are greater than or equal to A[F]. In other words, the relation (STRIP (BAGA A 1 F−1)) ⩽ A[F] ⩽ (STRIP (BAGA A F+1 NN)) should hold when the program halts. For instance, if F is NN÷2, then A[F] is the median of the array. The function is useful in computing percentiles and is fairly complex.

Hoare remarks that a sorting program would achieve the same purpose but would usually require much more time; the conditions for FIND are much weaker in that, for example, the elements to the left of F need not be sorted themselves, as long as none of them are greater than A[F].

The ALGOL representation of FIND is as follows:

```
FIND (F,NN,A); INTEGER F,NN; INTEGER ARRAY A[1:NN]
BEGIN
  INTEGER M,N;
  M ← 1;
  N ← NN;
  WHILE M < N DO
    BEGIN INTEGER R,I,J;
      R ← A[F];
      I ← M;
      J ← N;
      WHILE I ⩽ J DO
```

```
BEGIN WHILE A[I] < R DO I ← I+1;
  WHILE R < A[J] DO J ← J−1;
  IF I ≤ J THEN
  BEGIN EXCHANGE (A I J);
    I ← I+1;
    J ← J−1
  END
  END
  IF F ≤ J THEN N ← J
  ELSE IF I ≤ F THEN M ← I
        ELSE GO TO L
  END
  L:
END
```

The general strategy of the program FIND is to move "small" elements to the left and "large" elements to the right. These relative size categories are defined as being less than or not less than an arbitrary array element. The algorithm scans the array from left to right looking for a large element; when it finds one, it scans from right to left looking for a small element. When it finds one, it exchanges the large element and the small element it has already found, and the scan from the left continues where it left off until the next large element is found, and so on. When the scan from the left and the scan from the right meet somewhere in the middle, they define a split in the array. We can then show that all the elements to the left of the split are small and all those to the right are large.

The index F can be either to the left or to the right of the split, but suppose it is to the left. Then the elements to the right of the split can remain where they are; they are the largest elements in the array, and the element that will ultimately be in position F is to the left of the split. We then disregard the right portion of the array and repeat the process with the split as the upper bound of the array and with a refined definition of "large" and "small". We will eventually find a new split; suppose this split is to the left of F. We can then leave in place the elements of the array to the left of the split and work only with the elements to the right; we readjust the left bound of the array to occur at the split, and we repeat the process. Thus, the left and right bounds of the array move closer and closer together, but they always have F between them. Finally, they meet at F, and the algorithm halts.

The flowchart in Figure 1.6 follows Hoare's algorithm closely.

In this program, I is the pointer for the left-to-right scan, J is the pointer for the right-to-left scan, M and N are the lower and upper bounds of the "middle" portion of the array, and R is the value used to discriminate between small and large array elements. Hoare (1971) provided an informal manual proof of the correctness of his program. Deutsch (1973) and Igarashi, London, and Luckham (1975) have produced machine proofs. The proof we obtained required a minimal number (three) of intermediate assertions; however, one of the verification conditions produced was quite difficult to prove. This condition corresponds to the statement that the elements to the right of the right boundary dominate the elements to its left after an exchange is performed and a new right boundary is established. We present a sketch of this proof below.

4.4.1. Assertions for FIND

The input assertion q_s for FIND is (the conjunction of)

$$1 \leqslant F \leqslant NN,$$

$$A = AP,$$

The array AP is the initial version of A; we define it in the input assertion so that we can refer to it after we have modified A.

The output assertion q_H is

$$(STRIP\ (BAGA\ A\ 1\ F{-}1)) \leqslant A[F] \leqslant (STRIP\ (BAGA\ A\ F{+}1\ \ NN))$$
$$(BAGA\ A\ 1\ NN) = (BAGA\ AP\ 1\ NN).$$

The second conjunct of q_H states that when the program terminates, the array A is indeed a permutation of the initial array AP.

The intermediate assertion q_1 is

$$1 \leqslant M \leqslant F \leqslant N \leqslant NN$$
$$(STRIP\ (BAGA\ A\ 1\ M{-}1)) \leqslant (STRIP\ (BAGA\ A\ M\ NN))$$
$$(STRIP\ (BAGA\ A\ 1\ N)) \leqslant STRIP\ (BAGA\ A\ N{+}1\ NN))$$
$$(BAGA\ A\ 1\ NN) = (BAGA\ AP\ 1\ NN).$$

This assertion is reached whenever a new bound on the middle section of the array is established.

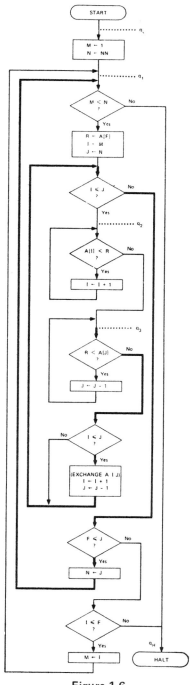

Figure 1.6.

The assertion q_2 is

$1 \leqslant M \leqslant F \leqslant N \leqslant NN$
(STRIP (BAGA A 1 M$-$1)) $\leqslant$ (STRIP (BAGA A M NN))
(STRIP (BAGA A 1 N)) $\leqslant$ (STRIP (BAGA A N+1 NN))
$M \leqslant I$
$J \leqslant N$
(STRIP (BAGA A 1 I$-$1)) $\leqslant R \leqslant$ (STRIP (BAGA A J+1 NN))
(BAGA A 1 NN) = (BAGA AP 1 NN).

The assertion q_3 is the same as the assertion q_2, with the additional conjunct

$$R \leqslant A[I].$$

4.4.2 The Proof

All but one of the verification conditions for this program were proved fairly easily. The one difficult condition corresponds to the path beginning at q_3 that follows the heavy line and finally ends at q_1. The verification-condition generator supplied us with the following hypotheses:

$1 \leqslant M \leqslant F \leqslant N \leqslant NN$,	(1.43)
(STRIP (BAGA A 1 M$-$1)) $\leqslant$ (STRIP (BAGA A M NN)),	(1.44)
(STRIP (BAGA A 1 N)) $\leqslant$ (STRIP (BAGA A N+1 NN)),	(1.45)
$M \leqslant I$,	(1.46)
$J \leqslant N$,	(1.47)
(STRIP (BAGA A 1 I$-$1)) $\leqslant R \leqslant$ (STRIP (BAGA A J+1 NN)),	(1.48)
$R \leqslant A[I]$,	(1.49)
(BAGA AP 1 NN) = (BAGA A 1 NN),	(1.50)
$\neg(R < A[J])$,	(1.51)
$I \leqslant J$,	(1.52)
$\neg(I+1 \leqslant J-1)$,	(1.53)
$F \leqslant J-1$.	(1.54)

The interesting consequence for this path is

(STRIP (BAGA A$'$ 1 J$-$1)) $\leqslant$ (STRIP (BAGA A$'$ (J$-$1)+1 NN)),

$$(1.55)$$

where

$$A' = (EXCHANGE\ A\ I\ J),$$

the array that results when elements A[I] and A[J] are interchanged in A.

The proof sketched below roughly parallels the proof produced by the inference system. Portions of the trace are shown in Section 7.

The (J−1)+1 term in the goal (1.55) is simplified to J, giving the goal

$$(STRIP\ (BAGA\ A'\ 1\ J-1)) \leqslant (STRIP\ (BAGA\ A'\ J\ NN)). \qquad (1.56)$$

The difficulty in the proof arises from the uncertainty about whether $J \leqslant I$. We are reasoning about an array segment, and it is not clear whether that segment is affected by the exchange or not. Hand analysis of the hypotheses (1.52) and (1.53) reveals that I = J or I = J−1. The value of a term like (BAGA (EXCHANGE A I J) 1 J−1) depends on which possibility is actually the case.

The system "simplifies" the term into

```
(IF J ≤ I THEN (BAGA A 1 J−1)
      ELSE (BAG  (STRIP (BAGA A 1 I−1))
                 A[J]
                 (STRIP (BAGA A I+1 J−1)))).
```

Intuitively, if $J \leqslant I$, then both I and J are outside the bounds of the array segment, whereas if $I < J$, then the array segment loses A[I] but gains A[J].

Similarly, the term

```
(BAGA (EXCHANGE A I J) J NN)
```

is "simplified" into

```
(IF J ≤ I  THEN (BAGA A J NN)
      ELSE (BAG (STRIP (BAGA A J J−1))
                A[I]
                (STRIP (BAGA A J+1 NN)))).
```

Note that (BAGA A J J−1) is empty; the ELSE clause is then

```
(BAG A[I] (STRIP (BAGA A J+1  NN))).
```

Our goal can thus be reduced to showing that

(IF J ≤ I THEN (STRIP (BAGA A 1 J−1))
        ELSE (STRIP (BAG (STRIP (BAGA A 1 I−1))
                          A[J]
                          (STRIP (BAGA A I+J J−1)))))

≤

(IF J ≤ I THEN (STRIP (BAGA A J NN))
        ELSE (STRIP (BAG A[I]
                          (STRIP (BAGA A J+1 NN)))))).

$$(1.57)$$

The system approaches the conditional expression by creating two contexts: In one context, J ≤ I holds, and in the other, I < J. In the first context we must prove that

(STRIP (BAGA A 1 J−1)) ≤ (STRIP (BAGA A J NN)). (1.58)

In the second context, the statement to be proved is

(STRIP (BAG (STRIP (BAGA A 1 I−1))
            A[J]
            (STRIP (BAGA A I+1 J−1))))

≤

(STRIP (BAG A[I]
            (STRIP (BAGA A J+1 NN)))). (1.59)

Note that in the first context, J = I by (1.52). In working on (1.58), (BAGA A J NN) is expanded to (BAG A[J] (STRIP (BAGA A J+1 NN))). Thus, (1.58) breaks into two subgoals:

(STRIP (BAGA A 1 J−1)) ≤ A[J] (1.60)

and

(STRIP (BAGA A 1 J−1)) ≤ (STRIP (BAGA A J+1 NN)).

$$(1.61)$$

Since I = J, (1.60) follows from (1.48) and (1.49), and (1.61) follows from (1.48) alone.

Work on the goal (1.59) proceeds in the second context, in which I < J. Since J−1 < I+1 (1.53), we know (BAGA A I+1 J−1) is empty.

The inequality (1.59) may thus be broken into four inequalities:

$$\text{(STRIP (BAGA A 1 I}-1\text{))} \leqslant A[I], \tag{1.62}$$

$$\text{(STRIP (BAGA A 1 I}-1\text{))} \leqslant \text{(STRIP (BAGA A J}+1 \text{ NN))}, \tag{1.63}$$

$$A[J] \leqslant A[I], \tag{1.64}$$

and

$$A[J] \leqslant \text{(STRIP (BAGA A J}+1 \text{ NN))}. \tag{1.65}$$

The goal (1.62) follows from the hypotheses (1.48) and (1.49). The goal (1.63) follows from (1.48). The goal (1.64) follows from (1.49) and (1.51). The goal (1.65) follows from (1.51) and (1.48). This completes the proof.

This proof is the most complex achieved by our deductive system so far.

5. Conclusion

5.1 Summary of Results

Complete proofs have been found of the correctness of the following algorithms:

(1) Finding the largest element of an array.
(2) Finding the quotient of two real numbers.
(3) Hoare's FIND program.
(4) The Euclidean algorithm for finding the greatest common divisor.
(5) The exponentiation program from King's thesis.
(6) Integer quotient and remainder.
(7) Integer multiplication by repeated addition.
(8) Computing the factorial of a nonnegative integer.

Theorems have been proved about the following algorithms:

(1) The pattern matcher.
(2) Unification.

(3) Exchanging two array elements (the theorem is that the bag of the contents of the array is unchanged).

(4) King's exchange sort.

We believe the system now has the power to do all of King's problem set except the linear inequalities problem, which is not really a proof about an algorithm.

5.2. Future Plans

We are currently applying the verifier to more and more complex programs in a variety of subject domains. We are continuously being forced to add new rules and occasionally to generalize old ones; a special-purpose rule that worked for one problem may not work for the next.

The deductive system is implemented in the QA4 language. Although QA4 is ideally suited for expressing our rules, it is an experimental system evaluated by an interpreter which is written in LISP; furthermore, it uses space inefficiently. Reboh and Sacerdoti have integrated QA4 into INTERLISP to produce a system known as QLISP (Wilber, 1976). QLISP programs are translated into LISP programs that can be evaluated by the LISP interpreter or even compiled. Furthermore, QLIP is much more conservative in its use of space. The QLISP system is considerably faster and more compact that the QA4 system. Our deductive system has already been translated into QLISP, and the same proofs are carried out many times faster.

QA4 subtly encourages its users to write depth-first search strategies, since it implements the goal mechanism by means of backtracking. The deductive system uses depth-first search and, for the most part, this has been the proper thing to do. There have been times, however, when we have felt the need for something more discriminating. Suppose, for example, we are trying to prove an expression of the form $x = y$. We can do this by trying to simplify x and then proving that the simplified x is equal to y, or we can try to find some assertion $a = b$ and prove $x = a$ and $y = b$. In the current system, we must exhaust one possibility before trying another, whereas we would like to be able to switch back and forth between different approaches, giving more attention to the one that currently seems to be making the best progress. In other words, we hope to use processes rather than backtracking in the implementation of the goal mechanism.

Finally, we hope to apply this work to the generation of counterexamples for "wrong" programs, to the generation of Floyd assertions, and to the automatic construction of programs. It seems inevitable that if we know how to reason about programs, that reasoning should be able to help us in the process of forming or changing a program. Rather than taking a handwritten and handdebugged program to a verifier for approval, we hope to collaborate with a system that will play an active role in the creation of the algorithm.

6. Appendix: Annotated Listing of the Deductive System

The deductive system has the overall structure shown in Figure 1.7. The names on the chart are either function names or goal classes. Only important substructures and relationships are included.

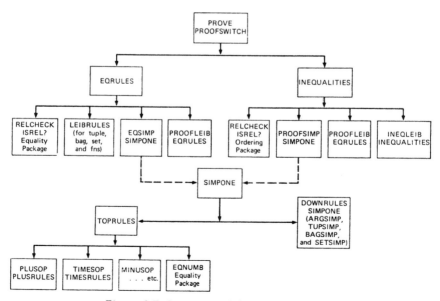

Figure 1.7. Structure of the Deductive System.

An annotated listing of the programs used for reasoning is presented below. An index of functions and goal classes is included at the end of this appendix. The reader will note how little of the space is devoted to general strategies and how much is devoted to subject-specific knowledge. Some of the programs use QA4 features that are not described in this paper. The reader can rely on the English explication of the programs, or he can refer to the QA4 manual (Rulifson et al., 1972).

To start a deduction, we say to the system

$$(GOAL \ \$PROVE \ \langle some \ statement \rangle)$$

- PROVE is a goal class: [4]

$$(TUPLE \ ANDSPLIT \ ORSPLIT \ ORSPLITMANY \ PROOFSWITCH)$$

- The rule ANDSPLIT takes a goal that is a conjunction of two or more expressions [5] and tries to prove each conjunct independently.

```
ANDSPLIT=(LAMBDA (AND ←X ←←Y)
              (ATTEMPT (GOAL $GOALCLASS $X)
              THEN
              (ATTEMPT (GOAL $GOALCLASS
                                    (AND $$Y))
                      ELSE
                      (FAIL))
              ELSE
              (FAIL] 6−8
```

If repeated applications of ANDSPLIT are successful eventually, the goal (AND)

---

[4] Bullets are used to indicate the beginning of the description of a new function.

[5] Variables with double prefixes, "←←" or "$$." respectively match or evaluate to a sequence of terms rather than a single term. In the rule, for example, Y can be bound to a set of terms, including the empty set.

[6] The right bracket represents a string of right parentheses long enough to balance the expression.

[7] The ATTEMPT statement is a conditional expression that tests for failure rather than falsehood, and has the additional power to restrict and control the effects of backtracking. For instance, suppose an expression of the form (ATTEMPT P THEN Q ELSE R) is being evaluated. If the evaluation of P is completed successfully, then Q is evaluated. On the other hand, if the evaluation of P results in a failure, then R is evaluated. A failure in P cannot cause the ATTEMPT statement to fail. A failure in Q or R, however, will cause the ATTEMPT statement to fail as usual.

[8] The ANDSPLIT rule occurs in several goal classes. The variable GOALCLASS that occurs in ANDSPLIT is bound by the system to whatever goal class was in effect when ANDSPLIT was invoked. Thus, ANDSPLIT applies to each conjunct the same goal class that was applied to the entire conjunction.

will be generated. However, (AND) is an assertion in the data base, and so the rule will then succeed.

● ORSPLIT applies to a goal that is the disjunction of two expressions and works on each separately.

```
ORSPLIT=(LAMBDA (OR ←X ←Y)
                (ATTEMPT (GOAL $GOALCLASS $X)
                    ELSE
                    (GOAL $GOALCLASS $Y]
```

The expression x is attempted as a goal first; if this is successful, we are done. Otherwise, ORSPLIT works on y; if it is unsuccessful, then a failure is generated.

● ORSPLITMANY is similar to ORSPLIT, except that it takes as a goal the disjunction of three or more expressions:

```
ORSPLITMANY =
    (LAMBDA (OR ←X ←Y ←Z ←←W)
                (ATTEMPT (GOAL $GOALCLASS $X)
                    ELSE
                    (GOAL $GOALCLASS (OR $Y $Z $$W))
```

The expression x is attempted first; if the proof is successful, the disjunction is true. Otherwise, the disjunction of the remaining expressions is established as a new goal. Continued failure to prove members of a disjunction will eventually cause ORSPLIT to be invoked.

● PROOFSWITCH attempts to apply the appropriate goal class to prove a goal. It determines whether the goal is an equality; if not, it is assumed to be an inequality. (Other goal classes could be added with little difficulty.) If the proof is successful, the goal is added to the data base as an assertion.

```
PROOFSWITCH=
        (LAMBDA (←F ←X)
                (PROG (DECLARE)
                    (IF (EQUAL $F (QUOTE EQ))
                        THEN
                        (GOAL $EQRULES ($F $X))
                        ELSE
                        (GOAL $INEQUALITIES ($F $X)))
                    (ASSERT ($F $X))
                    (RETURN ($F $X]
```

In either case, the appropriate set of rules is applied.

6.1. Equalities

● The equality class is

EQRULES =
(TUPLE ANDSPLIT ORSPLIT ORSPLITMANY RELCHECK EQTIMESDIVIDE
        EQSUBST LEIBT LEIBF LEIBB LEIBS EQSIMP PROOFLEIB)

● The rule RELCHECK merely checks the property lists of the expressions to
see if they are already known to be equal:

$$\text{RELCHECK}=(\text{LAMBDA} \leftarrow X \ (\text{ISREL? } \$X]$$

When RELCHECK is applied, x is bound to an equality statement, which is fed
to the ISREL? statement. ISREL? will succeed not only if the equality has been
explicitly asserted, but also if the equality follows by the transitive law from
other equalities or inequalities. ISREL? is the mechanism for making queries
about special relations. It will work with inequality relations, such as LT, GTQ,
and NEQ, as well as EQ.

EQTIMESDIVIDE and EQSUBST are rules for reasoning about numbers and
substitutions, respectively. They are discussed in the relevant sections.

● To prove $f(x) = f(y)$, try to prove $x = y$: this form of Leibniz's law for function applications is expressed by the rule LEIBF. The analogous rules for tuples,
sets, and bags are expressed by LEIBT, LEIBS, and LEIBB respectively.

```
LEIBF=(LAMBDA (EQ (←F ←X)
                  (←F ←Y))
              (PROG (DECLARE)
                    ($ASK ('(EQ $X $Y))
                          PROVE?)
                    (GOAL $EQRULES (EQ $X $Y]

LEIBT=(LAMBDA (EQ (TUPLE ←X ←←Z)
                  (TUPLE ←Y ←←W))
              (PROG (DECLARE)
                    (GOAL $EQRULES (EQ $X $Y))
                    (GOAL $EQRULES (EQ $Z $W]

LEIBS=(LAMBDA (EQ (SET ← X ←←Z)
                  (SET ←Y ←←Z))
              (GOAL $EQRULES (EQ $X $Y]

LEIBB=(LAMBDA (EQ (BAG ←X ←←Z)
                  (BAG ←Y ←←Z))
              (GOAL $EQRULES (EQ $X $Y]
```

The LEIBF rule asks the user if he wants that rule to be applied. The function ASK that performs the interaction is described in the section on utility functions.

• EQSIMP and PROOFLEIB are very time-consuming but also very powerful. EQSIMP says: to prove $x = y$, simplify x and try to prove that the simplified x is equal to y.

```
EQSIMP=(LAMBDA (EQ ←X ←Y)
            (PROG (DECLARE)
                  (SETQ ←X ($SIMPONE $X))
                  (GOAL $EQRULES (EQ $X $Y)))
            (BACKTRACK]
```

Since the program uses the BACKTRACK option, and since EQ implicitly takes a set as its argument, EQSIMP can work on y as well as on x. In other words, if it fails to simplify x, it will go ahead and try to simplify y.

• PROOFLEIB tries to make use of information stored in the data base. It is used to prove inequalities as well as equalities.

```
PROOFLEIB=(LAMBDA (←F ←X)
               (PROG (DECLARE)
                     (EXISTS ($F ←Y))
                     ($ASK ('(EQ $X $Y))
                          PROVE?)
                     (GOAL $EQRULES (EQ $X $Y]
```

It says: to prove $u = v$, find an assertion of the form $a = b$ and prove $u = a$ and $v = b$. The rule relies on user interaction to cut off bad paths. Note that if f is EQ, we can expect x and y to be sets, so that LEIBS will ultimately be called to prove the equality expression generated by PROOFLEIB.

The EXISTS statement searches the data base for an assertion that matches its argument, binding the ← variables appropriately. Subsequent failure causes it to look for another matching assertion. When no more assertions match, the EXISTS statement itself fails.

6.2. Inequalities

We now turn to the rules for proving inequalities.

```
INEQUALITIES=
(TUPLE ANDSPLIT RELCHECK ORSPLIT ORSPLITMANY PROOFSIMP
          INEQIFTHENELSE INEQSTRIPBAG INEQSTRIPSTRIP
          INEQSTRIPTRAN GTQLTQ LTQMANY FSUBSTRACT1
          FSUBSTRACT2 INEQTIMESDIVIDE EQINEQMONOTONE LTQPLUS
          PROOFLEIB INEQLEIB)
```

RELCHECK has been mentioned above.

● GTQLTQ says: to prove $y \geqslant x$, try to prove $x \leqslant y$:

        GTQLTQ =(LAMBDA (GTQ ←Y ←X)
                       (GOAL $INEQUALITIES (LTQ $X $Y]

● LTQMANY takes an inequality goal, such as

$$x_1 \leqslant x_2 \leqslant \cdots \leqslant x_n,$$

and breaks it into separate goals,

$$x_1 \leqslant x_2 \quad \text{and} \quad x_2 \leqslant x_3 \quad \text{and} \quad \cdots \quad \text{and} \quad x_{n-1} \leqslant x_n.$$

     LTQMANY=(LAMBDA (LTQ ←X ←Y ←Z ←←W)
                  (PROG (DECLARE)
                      (GOAL $INEQUALITIES (LTQ $X $Y))
                      (GOAL $INEQUALITIES
                            (LTQ $Y $Z $$W]

LTQPLUS, FSUBTRACT1, and FSUBTRACT2 are special rules for reasoning about numbers and are discussed in the relevant section.

● PROOFSIMP proves an expression $f(y)$ by trying to simplify y and proving the simplified expression.

     PROOFSIMP=(LAMBDA (PAND ←X (←F ←Y))
                    (PROG (DECLARE GOALCLASS1)
                      (SETQ ←GOALCLASS1 $GOALCLASS)
                      (ATTEMPT (SETQ ←X ($ARGSIMP $X))
                            ELSE
                            (FAIL))
                      (GOAL $GOALCLASS1 $X]

It has more general application than just to inequalities, although so far we have used it only for inequalities.

● INEQLEIB is similar to PROOFLEIB, but it works only for inequalities.

     INEQLEIB=(LAMBDA (←L ←X ←Y)
                  (PROG (DECLARE LOWER UPPER)
                      (EXISTS ($L ←LOWER ←UPPER))
                      ($ASK PROVE (' (LTQ $X $LOWER))
                        AND
                        (' (LTQ $UPPER $Y))
                        ?)
                      (GOAL $INEQUALITIES
                          (AND (LTQ $X $LOWER)
                              (LTQ $UPPER $Y]

L is expected to be LT or LTQ. To prove $x \lessgtr y$, for example, find an asserted statement LOWER < UPPER and prove $x \leqslant$ LOWER and UPPER $\leqslant y$.

- INEQIFTHENELSE is a rule that sets up a case analysis:

```
INEQIFTHENELSE=(LAMBDA
            (←F ←←W1 (IFTHENELSE ←X ←Y ←Z) ←←W2)
            (PROG (DECLARE VERICON)
                (ATTEMPT (SETQ ←VERICON
                                (CONTEXT PUSH LOCAL))
                        (ASSERT $X WRT $VERICON)
                        THEN
                        (GOAL $INEQUALITIES
                                ($F $$W1 $Y $$W2)
                                WRT $VERICON))
                (ATTEMPT (SETQ ←VERICON
                                (CONTEXT PUSH LOCAL))
                        (DENY $X WRT $VERICON)
                        THEN
                        (GOAL $INEQUALITIES
                                ($F $$W1 $Z $$W2)
                                WRT $VERICON)
                ELSE
                (RETURN (SUCCESS WITH
                                INEQIFTHENELSE]
```

For example, suppose the goal is (IF x THEN y ELSE z) $\leqslant w$. This rule establishes two subcontexts of the local context. In one of these contexts, x is true; in the other, x is false. In the first context, the rule tries to prove $y \leqslant w$, whereas in the second, it tries to prove $z \leqslant w$. Note that the subsystem that stores equalities and inequalities will cause a failure if an assertion (or a denial) would lead it to contradict what it knows. In that case the goal is considered to be achieved.

- INEQSTRIPBAG is an inequality rule that has a bag as one of its arguments.

```
INEQSTRIPBAG=(LAMBDA (←F ←←W (STRIP (BAG ←X ←←Y))
                        ←←Z)
                (GOAL $INEQUALITIES
                        (AND ($F $$W $X $$Z)
                                ($F $$W (STRIP (BAG $$Y))
                                $$Z ]
```

This rule would be invoked when we want to show, e.g., $w_1 \leqslant$ (STRIP (BAG c_1 c_2, ...)) $\leqslant w_2$. The intention here is to demonstrate that $w_1 \leqslant c_1 \leqslant w_2$ and $w_1 \leqslant c_2 \leqslant w_2$, and so forth. Ultimately, we might have to demonstrate that $w_1 \leqslant$ (STRIP (BAG)) $\leqslant w_2$. The special relations handler (ISREL?)

succeeds vacuously with any inequality relation where one of the arguments is (STRIP (BAG)).

6.3. Deduce

• DEDUCE is a goal class of rules that are guaranteed to terminate quickly. It is used when we want something more inquisitive than EXISTS but less time-consuming than PROVE, EQRULES, or INEQUALITIES.

```
DEDUCE=
(TUPLE RELCHECK ANDSPLIT ORSPLIT ORSPLITMANY LTPLUS
        FSUBTRACT1 FSUBTRACT2 LTQPLUS NOTATOM CONSTCAR
        CONSTCDR)
```

We have already described RELCHECK, ANDSPLIT, ORSPLIT, and ORSPLIT-MANY.

The other DEDUCE rules are for special applications and are discussed in the appropriate sections.

6.4. Simplification

• The top-level simplification function is SIMPONE. This function does not try to simplify its argument completely. It will find a partial simplification; repeated applications, if necessary, will completely simplify the expression.

The simplification rules may, of course, be added by the user. We expect that each simplification rule should make the expression simpler in some sense. Otherwise, the program may loop interminably.

```
SIMPONE=(LAMBDA ←GOAL1 (PROG
                (DECLARE SIMPGOAL)
                (IF (EQUAL (STYPE $GOAL1) NUMBER)
                   THEN
                   (FAIL))
                ($ASK $GOAL1 SIMPLIFY?)
                (SETQ ←SIMPGOAL
                    (ATTEMPT
                        (GOAL $TOPRULES $GOAL1)
                        ELSE
                        ($TRY $TOPRULES
                            (GOAL $DOWNRULES $GOAL1))))
                (PUT $GOAL1 SIMPLIFIED $SIMPGOAL
                    WRT ETERNAL)
                (RETURN $SIMPGOAL]
```

SIMPONE fails if it cannot simplify its argument at all. It treats numbers as being completely simplified. It asks the user for permission to go ahead. It tries a goal class, TOPRULES, on the expression.

- TOPRULES is a set of rules that work on the top level of the expression:

TOPRULES =
(TUPLE HASSIMP FAILINTODOWNRULES PLUSOP TIMESOP MINUSOP
   FIFTHENELSE BAGAOP SUBSTOP EXPZERO EXPEXP SUBPLUS
   SUBNUM GCDEQ ACCH CONSDIFF DIFDIF DIFFCONS DIFFONE
   MAXPLUS MAXONE BAGSTRIP ACCEX EQNUMB)

If any of these rules apply, SIMPONE returns the simplified expression as its value. Otherwise, it tries to simplify some subexpression of the given expression:

DOWNRULES=(TUPLE ARGSIMP TUPSIMP BAGSIMP SETSIMP)

ARGSIMP=(LAMBDA (←F ←X)
                 (SUBST ('($F $X))
                         (TUPLE $X ($SIMPONE $X]

TUPSIMP=(LAMBDA (TUPLE ←←X ←Y ←←Z)
                 (TUPLE $$X ($SIMPONE $Y)
                        $$Z)
                 BACKTRACK]

BAGSIMP=(LAMBDA (BAG ←X ←←Y)
                 (BAG ($SIMPONE $X)
                      $$Y)
                 BACKTRACK]

SETSIMP=(LAMBDA (SET ←X ←←Y)
                 (SET ($SIMPONE $X)
                      $$Y)
                 BACKTRACK]

The DOWNRULES simplify a complex expression by simplifying the component parts of the expression. If any of the DOWNRULES applies, SIMPONE applies the TOPRULES again to the new expression. SIMPONE calls the functions ASK and TRY that are described in the section on utility functions.

- SIMPONE puts the simplified expression on the property list of the original expression. In this way, if it ever comes across the original expression again, one of the TOPRULES, HASSIMP, will immediately know what simplification was found before.

HASSIMP=(LAMBDA ←X (IF (NOT (IN (SETQ ←X (GET $X SIMPLIFIED))
                                 (TUPLE DONE NOSUCHPROPERTY)))
                    THEN $X ELSE (FAIL]

● If the expression to be simplified is a set, tuple, or bag rather than a function application, none of the TOPRULES will apply to it. To avoid the cost of searching for a match among all the TOPRULES, the rule FAILINTODOWNRULES will first test for this condition and cause the entire goal statement to fail should it arise:

```
FAILINTODOWNRULES=(LAMBDA ←X
                        (IF (IN (STYPE $X)
                               (TUPLE TUPLE SET BAG))
                        THEN
                        (FAIL GOAL)
                        ELSE
                        (FAIL]
```

SIMPONE will then apply the DOWNRULES to the argument to see if any of its subexpressions can be simplified.

● One of the most general TOPRULES is EQNUMB, which replaces any expression by the "smallest" known equal expression:

```
EQNUMB=(LAMBDA ←X (PROG (DECLARE BEST EQSET)
                    (IF (EQUAL (SETQ ←EQSET (GET $X EQ))
                               NOSUCHPROPERTY)
                    THEN
                    (FAIL))
                    (SETQ ←BEST ($SHORTEST $EQSET))
                    (IF (EQUAL $BEST $X)
                        THEN
                        (FAIL)
                        ELSE
                        (RETURN $BEST]
```

The "smallest" element of a set is computed by the QA4 function SHORTEST, described among the utility functions. If EQNUMB fails to find a smaller representation for x, it fails.

● FIFTHENELSE=(LAMBDA (←F (IFTHENELSE ←W ←X ←Y))
                    ('(IFTHENELSE $W ($F $X)
                               ($F $Y])

FIFTHENELSE moves conditional expressions outside of function applications. An expression of the form

$$f(\text{IF } w \text{ THEN } x \text{ ELSE } y)$$

translates into

$$\text{IF } w \text{ THEN } f(x) \text{ ELSE } f(y).$$

The remaining rules in TOPRULES are discussed in the sections dealing with special subject domains.

6.5. Reasoning About Numbers

6.5.1. Equality and inequality rules

● EQTIMESDIVIDE is an EQRULE. It means that to prove $w = (x/y)*z$, one should prove $w*y = x*z$:

```
EQTIMESDIVIDE=(LAMBDA (EQ ←W (TIMES (DIVIDE ←X ←Y)
                                      ←←Z))
                     (GOAL $EQRULES (EQ (TIMES $Y $W)
                                        (TIMES $X $$Z)))
              BACKTRACK]
```

Some inequality rules that know about numbers are presented below.

● LTQPLUS says that to prove $i \leqslant j+k$, one should prove $i \leqslant j$ and $0 \leqslant k$:

```
LTQPLUS=(LAMBDA (LTQ ←I (PLUS ←J ←K))
               (GOAL $DEDUCE (AND (LTQ $I $J)
                                  (LTQ 0 $K)))
        BACKTRACK]
```

First, the rule attempts to prove that $i \leqslant j$ and $0 \leqslant k$. If either of these proofs is unsuccessful, then the backtracking mechanism will interchange the bindings of the arguments of LTQPLUS. This then leads to an attempt to prove $i \leqslant k$ and $0 \leqslant j$.

● LTPLUS is the analogue of LTQPLUS for LT:

```
LTPLUS=(LAMBDA (LT ←I (PLUS ←J ←K))
              (GOAL $DEDUCE (AND (LTQ $I $J)
                                 (LT 0 $K)))
       BACKTRACK]
```

It means: to prove $i < j+k$, prove $i \leqslant j$ and $0 < k$. It can backtrack to reverse the roles of j and k.

● FSUBTRACT1 and FSUBTRACT2 allow us to remove subtraction from the goal; for example, to prove $x-y \leqslant z$, try to prove $x \leqslant y+z$.

```
FSUBTRACT1=(LAMBDA (←F (SUBTRACT ←X ←Y)
                        ←Z)
                  (GOAL $GOALCLASS ($F $X (PLUS $Y $Z]
```

```
FSUBTRACT2=(LAMBDA (←F ←X (SUBTRACT ←Y ←Z))
                   (GOAL $GOALCLASS ($F (PLUS $X $Z)
                                       $Y]
```

- EQINEQMONOTONE says: to prove $w+x \leq y+z$, prove $w \leq y$ and $x \leq z$ or $w \leq z$ and $x \leq y$.

```
EQINEQMONOTONE=(LAMBDA (←L (PLUS ←W ←X)
                           (PLUS ←Y ←Z))
               (PROG (DECLARE)
                    ($ASK PROVE ('($L $W $Y))
                          AND
                          ('($L $X $Z))
                          ?)
                    (GOAL $GOALCLASS
                          (AND ($L $W $Y)
                               ($L $X $Z]
               BACKTRACK]
```

- The rule INEQTIMESDIVIDE is similar to EQTIMESDIVIDE except that it must check that the denominator is nonnegative before multiplying out:

```
INEQTIMESDIVIDE=(LAMBDA (←F ←W (TIMES (DIVIDE ←X ←Y)
                                       ←←Z))
                (PROG (DECLARE)
                     (GOAL $DEDUCE (LT 0 $Y))
                     (GOAL $INEQUALITIES
                           ($F (TIMES $Y $W)
                               (TIMES $X $$Z))))
                BACKTRACK]
```

This rule says: to prove $w < (x/y)*z$, say, in the case that $0 < y$, try to prove $w*y < x*z$.

6.5.2. Numerical demons

- When $x \geq y$ is asserted, assert that $y \leq x$:

```
(WHEN EXP (GTQ ←X ←Y)
     INDICATOR MODELVALUE THEN (ASSERT (LTQ $Y $X)
                                       WRT $VERICON]
```

These demons make their assertions with respect to the current context, VERICON.

- Whenever $x+y \leq x+z$ is asserted, we want to conclude that $y \leq z$:

```
(WHEN EXP (LTQ (PLUS ←X ←Y)
              (PLUS ←X ←Z))
     INDICATOR MODELVALUE THEN
     (ASSERT (LTQ $Y $Z)
             WRT $VERICON]
```

• Whenever $w-x \leqslant y$ is asserted, assert $w \leqslant x+y$, simplifying the right side if possible.

```
(WHEN EXP (LTQ (SUBTRACT ←W ←X)
               ←Y)
     INDICATOR MODELVALUE THEN
     (PROG (DECLARE RTSIDE)
           (SETQ ←RTSIDE
                 ($TRYALL $PLUSRULES
                          (' (PLUS $Y $X]
           (ASSERT (LTQ $W $RTSIDE)
                   WRT $VERICON))))
```

• Whenever $(w-x)+x \leqslant y$ is asserted, then assert $w \leqslant y$:

```
WHEN EXP (LTQ (PLUS (SUBTRACT ←W ←X)
                    ←X)
              ←Y)
     INDICATOR MODELVALUE THEN
     (ASSERT (LTQ $W $Y)
             WRT $VERICON]
```

Certain demons are intended exclusively for the integer domain.

• $x < y \supset x+1 \leqslant y$:

```
(WHEN EXP (LT ←X ←Y)
     INDICATOR MODELVALUE THEN
     (ASSERT (LTQ (PLUS $X 1)
                  $Y)
             WRT $VERICON]
```

• $x > y \supset y+1 \leqslant x$:

```
(WHEN EXP (GT ←X ←Y)
     INDICATOR MODELVALUE THEN
     (ASSERT (LTQ (PLUS $Y 1)
                  $X)
             WRT $VERICON]
```

Whenever $w-x < y$ is denied, *deny* $w \leqslant y+x-1$, simplifying if possible:

```
(WHEN EXP (LT (SUBTRACT ←W ←X)
              ←Y)
     INDICATOR MODELVALUE PUTS FALSE THEN
     (PROG (DECLARE RTSIDE)
           (SETQ
                 ←RTSIDE
```

```
($TRYALL $PLUSRULES
    ('(PLUS $Y $X
          (MINUS 1))))
(DENY (LTQ $W $RTSIDE)
WRT $VERICON]
```

6.5.3. Numerical simplification

Much of the knowledge the system has about numbers is embedded in the simplifier. For efficiency, these rules have been arranged hierarchically. For example, only one rule, PLUSOP, in TOPRULES deals with sums.[9]

```
PLUSOP=(LAMBDA (PAND ←Y (PLUS ←←X))
                ($TRYALLFAIL $PLUSRULES $Y]
```

However, this one rule coordinates a multitude of other rules. All the rules that operate on plus expressions are in the goal class PLUSRULES.

```
PLUSRULES=(TUPLE PLUSEMPTY PLUSSINGLE PLUSZERO PLUSPLUS
                 PLUSMINUS PLUSDIFFERENCE PLUSCOMBINE
                 PLUSNUMBER)
```

The strategy PLUSOP uses is to apply all the PLUSRULES to its argument until no further simplification is possible. (The function TRYALLFAIL, that expresses this strategy, is described among the utility functions.) If PLUSOP can find no simplification at all, it fails.

Most of the PLUSRULES are quite simple.

● The sum of the empty bag is 0:

$$PLUSEMPTY=(LAMBDA (PLUS) 0)$$

● The sum of a bag of one element is that element itself:

$$PLUSSINGLE=(LAMBDA (PLUS ←X) \$X]$$

(i.e., $+x = x$).

● $x+0 = +x$:

$$PLUSZERO=(LAMBDA (PLUS ←←X 0) (' (PLUS \$\$X]$$

---

[9] A pattern (PAND *pat*1 *pat*2) will match an argument if and only if *pat*1 and *pat*2 both match that argument.

Note that this rule implicitly says

$$0 + x = +x,$$

$$x + 0 + y = x + y,$$

$$x + y + 0 + z = x + y + z,$$

and so forth, because PLUS takes a bag as its argument.

- $((x_1+x_2+...)+y_1+y_2+...) = (x_1+x_2+...+y_1+y_2+...)$:

$$\text{PLUSPLUS=(LAMBDA (PLUS (PLUS} \leftarrow\leftarrow\text{X)} \leftarrow\leftarrow\text{Y)}$$
$$('\,\text{(PLUS \$\$X \$\$Y]}$$

- $x+(-x)+y = +y$:

$$\text{PLUSMINUS=(LAMBDA (PLUS} \leftarrow\text{X (MINUS} \leftarrow\text{X)} \leftarrow\leftarrow\text{Y)}$$
$$('\,\text{(PLUS \$\$Y]}$$

- $x+(y-z)+w = x+y+w+(-z)$:

$$\text{PLUSDIFFERENCE=(LAMBDA (PLUS} \leftarrow\text{X (SUBTRACT} \leftarrow\text{Y} \leftarrow\text{Z)} \leftarrow\leftarrow\text{W)}$$
$$(\$\text{TRY (TUPLE PLUSMINUS)}$$
$$('\,\text{(PLUS \$Y \$X \$\$W (MINUS \$Z]}$$

Note that PLUSDIFFERENCE recommends that PLUSMINUS be attempted immediately afterward. This is merely advice; if PLUSMINUS does not apply, nothing is lost. (TRY is described in the section on utility functions.)

- $x+x+y = 2*x+y$:

$$\text{PLUSCOMBINE=(LAMBDA (PLUS} \leftarrow\text{X} \leftarrow\text{X} \leftarrow\leftarrow\text{Y)}$$
$$(\$\text{TRYSUB \$TIMESRULES ON}$$
$$('\,\text{(TIMES 2 \$X))}$$
$$\text{IN}$$
$$('\,\text{(PLUS (TIMES 2 \$X) \$\$Y]}$$

Note that PLUSCOMBINE recommends that the $2*x$ term be simplified if possible. (TRYSUB is explained in the section on utility functions.)

- If two elements of a plus expression are syntactically numbers, PLUS-NUMBER will add them up:

$$\text{PLUSNUMBER=(LAMBDA (PLUS} \leftarrow\text{X} \leftarrow\text{Y} \leftarrow\leftarrow\text{Z)}$$
$$\text{(PROG (DECLARE SUM)}$$

                              ($INSIST (EQUAL (STYPE $X)
                                             NUMBER))
                              ($INSIST (EQUAL (STYPE $Y)
                                             NUMBER))
                              (SETQ ←SUM (PLUS $X $Y))
                              (RETURN (PLUS $SUM $$Z)))
                         BACKTRACK]

- The rule TIMESOP is strategically similar to PLUSOP:

         TIMESOP=(LAMBDA (PAND ←Y (TIMES ←←X))
                         ($TRYALLFAIL $TIMESRULES $Y]

It will apply all the TIMESRULES to the expression in question. TIMESRULES
is

TIMESRULES =
(TUPLE TIMESEMPTY TIMESSINGLE TIMESZERO TIMESONE TIMESPLUS
      TIMESTIMES CANCEL SQRULE TIMESEXP TIMESDIVIDEONE)

- The product of the empty bag is 1:

              TIMESEMPTY=(LAMBDA (TIMES) 1)

- The product of a bag of one element is that element itself:

         TIMESSINGLE=(LAMBDA (TIMES ←X) $X)

- $0*y = 0$:

         TIMESZERO=(LAMBDA (TIMES 0 ←←Y) 0)

- $1*x = x$:

              TIMESONE=(LAMBDA (TIMES 1 ←←X)
                              (' (TIMES $$X]

Recall that these rules also imply

$$x*1*y = x*y,$$

$$x*0*y*z = 0,$$

and so forth.

- $(x+y)*z = x*z+(+y)*z$ (distributive law), where $+y$ represents a sum of one or
more terms:

TIMESPLUS=(LAMBDA (TIMES (PLUS ←X ←←Y) ←←Z)
($TRY $PLUSRULES ($TRYSUB $PLUSRULES ON ('(PLUS $$Y))
                  IN
                  ('(PLUS (TIMES $X $$Z)
                          (TIMES (PLUS $$Y) $$Z]

(Some simplification is attempted immediately on $+y$ and on $x*z + (+y)*z$. TRYSUB is explained in the section on utility functions.)

- $((x_1*x_2*...)*y_1*y_2...) = (x_1*x_2*...*y_1*y_2*...)$:

TIMESTIMES=(LAMBDA (TIMES (TIMES ←←X) ←←Y)
           (' (TIMES $$X $$Y]

- $x*(1/y)*z = (x/y)*z$:

TIMESDIVIDEONE=(LAMBDA (TIMES ←X (DIVIDE 1 ←Y) ←←Z)
               (' (TIMES (DIVIDE $X $Y)
                         $$Z]

- $x*(y/x)*z = y*z$:

CANCEL=(LAMBDA (TIMES ←X (DIVIDE ←Y ←X) ←←Z)
       (' (TIMES $Y $$Z]

- $x*x*y = x^2* y$:

SQRULE=(LAMBDA (TIMES ←X ←X ←←Y)
       ($TRY (TUPLE TIMESSINGLE)
       (' (TIMES (EXP $X 2) $$Y]

- $x*x^n*y = x^{n+1}*y$:

TIMESEXP=(LAMBDA (TIMES ←X (EXP ←X ←N) ←←Y)
         ($TRYSUB $PLUSRULES ON ('(PLUS $N 1))
         IN
         (' (TIMES (EXP $X (PLUS $N 1)) $$Y]

- To the reader who has gotten this far, MINUSOP will be self-explanatory:

MINUSOP=
(LAMBDA (MINUS ←X)
        (GOAL (TUPLE MINUSZERO MINUSMINUS MINUSPLUS)
        (MINUS $X]

Note that MINUSOP, unlike PLUSOP and TIMESOP, does not apply *all* the rules to the expression, but will return the value of the first rule that does not fail.

- $-0 = 0$:

$$\text{MINUSZERO}=(\text{LAMBDA (MINUS 0) 0})$$

- $-(-x) = x$:

$$\text{MINUSMINUS}=(\text{LAMBDA (MINUS (MINUS} \leftarrow\text{X)) \$X})$$

- $-(x+y) = (-x)+(-y)$:

```
MINUSPLUS=(LAMBDA (MINUS (PLUS ←X ←←Y))
                ($TRY $PLUSRULES (PLUS (MINUS $X)
                                       (MINUS (PLUS $$Y]
```

At present there are only two subtraction rules, and so we do not combine them into one operator:

- $x-y = x+(-y)$:

```
SUBPLUS=(LAMBDA (SUBSTRACT ←X ←Y)
              ($TRY $PLUSRULES (' (PLUS $X (MINUS $Y]
```

- If x and y are both numbers and not variables, SUBNUM actually evaluates $x-y$:

```
SUBNUM=LAMBDA (SUBTRACT ←X ←Y)
            (PROG (DECLARE)
                  ($INSIST (AND (EQUAL (STYPE $X) NUMBER)
                                (EQUAL (STYPE $Y)
                                       NUMBER)))
                  (RETURN (= (SUBTRACT $X $Y]
```

The "=" sign forces the system to evaluate what it would otherwise merely instantiate. INSIST is another utility function. Two more rules about exponentiation are given below.

- $x^0 = 1$:

$$\text{EXPZERO}=(\text{LAMBDA (EXP} \leftarrow\text{X 0) 1 })$$

- $(x^y)^z = x^{y*z}$:

```
EXPEXP=(LAMBDA (EXP (EXP ←X ←Y) ←Z)
             ($TRYSUB $TIMESRULE ON (' (TIMES $Y $Z))
             IN
             (' (EXP $X (TIMES $Y $Z]
```

Note that **EXPEXP** recommends that the **TIMESRULES** be applied to the product $y*z$; this is heuristic advice that could have been omitted.

- $gcd(x\ x) = x$:

```
GCDEQ=(LAMBDA (GCD ←X ←Y)
        (PROG (DECLARE)
              (GOAL $DEDUCE (EQ ←X ←Y))
              (RETURN $X]
```

The GCD is the greatest common divisor.

6.6. Reasoning About Arrays

Most of the knowledge about arrays embedded in the system is expressed as simplification rules.

- (ACCESS (CHANGE A I T) I) = T,
  $I \neq J \supset$ (ACCESS (CHANGE A I T) J) = (ACCESS A J):

```
ACCH=(LAMBDA (ACCESS (CHANGE ←A ←I ←T)
                     ←J)
        (PROG (DECLARE)
              (ATTEMPT (GOAL $DEDUCE (EQ $I $J))
                       THEN
                       (RETURN $T))
              (GOAL $DEDUCE (NEQ $I $J))
              (RETURN (ACCESS $A $J]
```

ACCH is one of the TOPRULES, as are the rules below, ACCEX, MAXONE, MAX, and MAXPLUS.

- (EXCHANGE A I J) is a higher-level function whose output is the array A with the values of A[I] and A[J] exchanged. The value of (ACCESS (EXCHANGE A I J) K) depends on whether or not K equals I or J, i.e., whether the element here accessed was affected by the exchange. If K = I, the value is A[J]. If K = J, the value is A[I]. If K is neither I nor J, the value is the original value of A[K], since the location has not been affected by the exchange. The rule fails if it cannot be determined whether or not K= I or K=J. This information is embodied in the rule ACCEX:

```
ACCEX=(LAMBDA (ACCESS (EXCHANGE ←A ←I ←J)
                      ←K)
        (PROG (DECLARE)
              (ATTEMPT (GOAL $DEDUCE (EQ $K $I ))
                       THEN
                       (RETURN (ACCESS $A $J)))
```

```
                    (ATTEMPT (GOAL $DEDUCE (EQ $K $J))
                             THEN
                             (RETURN (ACCESS $A $I)))
                    (GOAL $DEDUCE (AND (NEQ $K $I)
                                       (NEQ $K $J)))
                    (RETURN (ACCESS $A $K]
```

● The maximum of an array, MAXA, is a function of three arguments: the array, the lower bound, and the upper bound.

$$(MAXA\ A\ I\ J)=(MAX\ A[I]\ ,\ A[I+1]\ ,\ ...,\ A[J]\ )$$

$$(MAXA\ A\ I\ I)=A[I]\ :$$

```
MAXONE=(LAMBDA (MAXA ←A ←I ←J)
               (PROG (DECLARE)
                     (GOAL $DEDUCE (EQ $I $J))
                     (RETURN (ACCESS $A $I]
```

● If

$$(MAXA\ A\ I\ J) \leqslant A[J+1]\ ,$$

then

$$(MAXA\ A\ I\ J+1) = A[J+1]\ .$$

On the other hand, if

$$(MAXA\ A\ I\ J)\ >\ A[J+1]\ ,$$

then

$$(MAXA\ A\ I\ J+1) = (MAXA\ A\ I\ J);$$

```
MAXPLUS=(LAMBDA
        (MAXA ←A ←I (PLUS ←J 1))
        (PROG (DECLARE)
              (ATTEMPT (GOAL $DEDUCE (LTQ (MAXA $A $I $J)
                                          (ACCESS $A (PLUS $J 1]
                       THEN
                       (RETURN (ACCESS $A (PLUS $J 1))))
              (GOAL $DEDUCE (LT (ACCESS $A (PLUS $J 1))
                                (MAXA $A $I $J))
              (RETURN (MAXA $A $I $J]
```

● Recall that (BAGA A I J) is (BAG A[I], A[I+1], ..., A[J]). Because of the crucial part this function plays in assertions about sortlike programs, we have many rules for it.

BAGARULES=
(TUPLE BAGAPLUS BAGAEMPTY BAGAII ARGSIMP BACH BAGEX BAGEX1
    BAGAMINUS BAGALOWERPLUS BAGEXCOMPLICATED)

- These rules are controlled by the rule BAGAOP, one of the TOPRULES:

$$BAGAOP=(LAMBDA\ (PAND \leftarrow Y\ (BAGA \leftarrow \leftarrow X))$$
$$(\$TRYALLFAIL\ \$BAGARULES\ \$Y]$$

Thus, the BAGARULES will be tried whenever we are simplifying an expression of the form (BAGA A I J).

- If $I \leqslant J+1$, then (BAGA A I J+1) = (BAG A[J+1] (STRIP (BAGA A I J))):

BAGAPLUS=(LAMBDA (BAGA ←A ←I (PLUS 1 ←J))
                (PROG (DECLARE)
                        (GOAL $DEDUCE (LTQ $I (PLUS 1 $J)))
                        (RETURN (BAG (ACCESS $A (PLUS $J 1))
                                        (STRIP (BAGA $A $I $J]

- If $I < J$, then (BAGA A J I) is the empty bag:

BAGAEMPTY=(LAMBDA (BAGA ←A ←J ←I)
                    (PROG (DECLARE)
                            (GOAL $DEDUCE (LT $I $J))
                            (RETURN (BAG]

- (BAGA A I I) is (BAG A[I]):

BAGAII=(LAMBDA (BAGA ←A ←I ←J)
                (PROG (DECLARE)
                        (GOAL $DEDUCE (EQ $I $J))
                        (RETURN (BAG (ACCESS $A $I]

- If $I \leqslant J$, then (BAGA A I J) = (BAG (STRIP (BAGA A I J−1)) A[J]):

BAGAMINUS=(LAMBDA (BAGA ←A ←I ←J)
        (PROG (DECLARE)
                ($INSIST (EQUAL (STYPE $J)
                                IDENT))
                (GOAL $DEDUCE (LTQ $I $J))
                (RETURN (BAG (ACCESS $A $J)
                                (STRIP (BAGA $A $I (SUBTRACT $J 1]

Since this rule would apply so often, it is restricted by forcing J to be an identifier rather than a complex expression.

● If L ⩽ M, then

(BAGA A L M) = (BAG A[L] (STRIP (BAGA A L+1 M))):

BAGALOWERPLUS=
(LAMBDA (BAGA ←ARNAME ←L ←M)
        (PROG (DECLARE F LOWER UPPER W)
              (EXISTS (←F ←←V (STRIP (BAGA $ARNAME ←LOWER
                                                ←UPPER))
                       ←←W))
              (GOAL $DEDUCE (EQ $LOWER (PLUS 1 $L)))
              (RETURN (BAG (ACCESS $ARNAME $L)
                           (STRIP (BAGA $ARNAME (PLUS 1 $L)
                                   $M]

This rule tries to determine if its application is desirable by checking in the
model for any relationship involving an array segment with lower bound equal
to L+1; if no such relationship exists, it is doubtful that the proposed simplification will lead to a proof.

● If I ⩽ J ⩽ K, then

(BAGA (CHANGE A J T) I K) =

(BAG T (STRIP (BAGA A I K))) ~ (BAG A[J]).

On the other hand, if J < I or K < J,

(BAGA (CHANGE A J T) I K) = (BAGA A I K).

(The notation ~ means the difference between two bags.) In other words,
making an assignment to an array element whose index is outside the bounds
of a segment does not affect the segment. However, if the index is within
bounds of the segment, then the corresponding bag will lose the old value of the
array element but gain the new value:

BACH=
(LAMBDA
  (BAGA (CHANGE ←A ←J ←T) ←I ←K)
  (PROG
    (DECLARE)
    (ATTEMPT
      (GOAL $DEDUCE (LTQ $I $J $K))
      THEN
      (RETURN
        (=
          ($TRY
           $DIFFRULES

```
($TRYSUB
 (TUPLE ACCH ACCEX)
 ON
 (' (ACCESS $A $J))
 IN
 ($TRYSUB $BAGARULES ON (' (BAGA $A $I $K))
          IN
          (' (DIFFERENCE (BAG $T
                                (STRIP (BAGA $A $I $K)))
                         (BAG (ACCESS $A $J)))))))))))
 (GOAL $DEDUCE (OR (LT $J $I) (LT $K $J)))
 (RETURN (BAGA $A $I $K]
```

The rule BACH contains many recommendations about possible future sim-
plifications. These recommendations are included to promote efficiency; the
simplifier would eventually try the recommended rules even if the advice were
omitted. The advice-giving functions TRY and TRYSUB are described in the
section on utility functions.

• As mentioned above, (EXCHANGE A I J) is the array A with the values of
A[I] and A[J] interchanged. If I and J are either both inside or both outside
an array segment, then the exchange operation has no effect on the bag corre-
ponding to that segment:

```
BAGEX = (LAMBDA (BAGA (EXCHANGE ←A ←I ←J) ←L ←M)
                (PROG (DECLARE)
                      (GOAL $DEDUCE (LTQ $I $J))
                      (ATTEMPT
                       (GOAL $DEDUCE
                             (OR (AND (LTQ $L $I)
                                      (LTQ $J $M))
                                 (LT $J $L)
                                 (LT $M $I)
                                 (AND (LT $I $L)
                                      (LT $M $J))))
                      THEN
                      (RETURN (BAGA $A $L $M))
                      ELSE
                      (FAIL]
```

For simplicity, BAGEX requires that I ≤ J.

• If elements A[I] and A[J] are exchanged, and if J is in the array segment and
I is not, or if I is in the segment and J is not, then the corresponding bag is
indeed affected by the exchange operation. For instance, in the case in which J

is in the segment and I is not, if the segment is bounded by L and M, the new
bag is

$$
\begin{array}{l}
\text{(BAG (STRIP (BAGA A L J–1)} \\
\qquad\text{A[I]} \\
\qquad\text{(STRIP (BAGA A J+1M))):}
\end{array}
$$

```
BAGEX1=
(LAMBDA
  (BAGA (EXCHANGE ←A ←I ←J) ←L ←M)
  (PROG (DECLARE)
        (GOAL $DEDUCE (LTQ $I  $J))
        (ATTEMPT (GOAL $DEDUCE (AND (LT $I $L)
                                   (LTQ $L $J)
                                   (LTQ $J $M)))
                THEN
                (RETURN (BAG (STRIP (BAGA $A $L (SUBTRACT $J 1)))
                             (ACCESS $A $I)
                             (STRIP (BAGA $A (PLUS 1 $J) $M))))
        (ATTEMPT (GOAL $DEDUCE (AND (LT $M $J)
                                   (LTQ $L $I)
                                   (LTQ $I $M)))
                THEN
                (RETURN (BAG (STRIP (BAGA $A $L (SUBTRACT $I 1)))
                             (ACCESS $A $J)
                             (STRIP (BAGA $A (PLUS 1 $I) $M))))
                ELSE
                (FAIL]
```

● BAGEXCOMPLICATED handles the case in which it can be determined
that one of the exchanged elements is within or outside the array segment, but
the location of the other array element is uncertain. The result is then a conditional expression. For example, if J is known to be outside the segment but
I is only known to be greater than or equal to the lower limit L, the result is

$$
\begin{array}{l}
\text{(IF M < I THEN (BAGA A L M)} \\
\qquad\text{ELSE (BAG (STRIP (BAGA A L I–1))} \\
\qquad\qquad\text{A[J]} \\
\qquad\qquad\text{(STRIP (BAGA A I+1 M))):}
\end{array}
$$

```
BAGEXCOMPLICATED=
(LAMBDA
  (BAGA (EXCHANGE ←A ←I ←J) ←L ←M)
  (PROG
```

```
(DECLARE)
(GOAL $DEDUCE (LTQ $I $J))
(ATTEMPT (GOAL $DEDUCE (AND (LTQ $L $I)
                            (LTQ $M $J)))
         THEN
         (RETURN (IFTHENELSE
                  (LT $M $I)
                  (BAGA $A $L $M)
                  (BAG (STRIP (BAGA $A $L (SUBTRACT $I 1)))
                       (ACCESS $A $J)
                       (STRIP (BAGA $A (PLUS 1 $I) $M)))))))
(ATTEMPT (GOAL $DEDUCE (AND (LTQ $J $M)
                            (LTQ $L $J)))
         THEN
         (RETURN (IFTHENELSE
                  (LTQ $L $I)
                  (BAGA $A $L $M)
                  (BAG (STRIP (BAGA $A $L (SUBTRACT $J 1)))
                       (ACCESS $A $I)
                       (STRIP (BAGA $A (PLUS 1 $J) $M)))))
         ELSE
         (FAIL]
```

BAGEXCOMPLICATED comes after BAGEX and BAGEX1 in the goal class BAGARULES because we prefer the definite answer they provide to the conditional expression returned by BAGEXCOMPLICATED.

All the rules in this section have been simplification rules. There also are two inequalities rules that pertain to arrays, INEQSTRIPTRAN and INEQSTRIP-STRIP.

• To prove that every element in an array segment is less than (or less than or equal to) some quantity C, find an array segment that properly contains the given segment such that every element in the larger segment is less than some element D that is, in turn, less than or equal to C:

```
INEQSTRIPTRAN=
(LAMBDA (←F (STRIP (BAGA ←ARNAME ←L ←M))←C)
        (PROG (DECLARE LOWER UPPER D)
              (EXISTS ($F (STRIP (BAGA $ARNAME ←LOWER
                                       ←UPPER)) ←D))
              (GOAL $DEDUCE (AND (LTQ $LOWER $L)
                                 (LTQ $M $UPPER)
                                 (LTQ $D $C]
```

• To prove some ordering relation < or ≤ between all the elements of two array segments, S_1 and S_2, find relations of the same sense involving S_1' and C, and involving D and S_2'. Then show that S_1' and S_2' contain S_1 and S_2 respectively, and that C is less than or equal to D.

```
INEQSTRIPSTRIP=
(LAMBDA (←F (STRIP (BAGA ←A ←I ←J))
                (STRIP (BAGA ←A ←K ←L)))
            (PROG (DECLARE LOWER1 UPPER1 LOWER2 UPPER2 C D)
                (ATTEMPT (EXISTS ($F (STRIP (BAGA $A ←LOWER1
                                                        ←UPPER1))
                                        ←C))
                        (EXISTS ($F ←D
                                        (STRIP (BAGA $A ←LOWER2
                                                        ←UPPER2))))
                        (GOAL $DEDUCE (AND (LTQ $LOWER1 $I)
                                        (LTQ $J $UPPER1)
                                        (LTQ $LOWER2 $K)
                                        (LTQ $L $UPPER2)
                                        (LTQ $C $D)))
                ELSE
                (FAIL]
```

6.7. Reasoning About Bags

We have accumulated a number of rules about bags. Many of these rules have set-theoretic counterparts, which could have been included, but we have needed only bags in our proofs.

We use the QA4 function DIFFERENCE to mean the difference between bags, written informally as "$\sim$".

- $(BAG\ x\ y)\sim(BAG\ x) = (BAG\ y)$:

```
DIFFXX=(LAMBDA (DIFFERENCE (BAG ←X ←←Y)
                                (BAG ←X))
            (BAG $$Y]
```

- $cons(x, y \sim z) = cons(x, y) \sim z$:

```
CONSDIFF=(LAMBDA (CONS ←X (DIFFERENCE ←Y ←←Z))
                ( (DIFFERENCE (CONS $X $Y)
                                $$Z]
```

- $(x \sim y) \sim z = x \sim y \sim z$:

```
DIFDIF=(LAMBDA (DIFFERENCE (DIFFERENCE ←X ←←Y)
                                ←←Z)
            ('(DIFFERENCE $X $$Y $$Z]
```

- $cons(x, y) \sim (BAG\ x) \sim u = y \sim u$:

$$DIFFCONS = (LAMBDA\ (DIFFERENCE\ (CONS \leftarrow X \leftarrow Y)$$
$$(BAG \leftarrow X)$$
$$\leftarrow \leftarrow U)$$
$$(\$TRY\ (TUPLE\ DIFFONE)$$
$$('\ (DIFFERENCE\ \$Y\ \$\$U]$$

- (DIFFERENCE x) is taken to be x itself:

$$DIFFONE = (LAMBDA\ (DIFFERENCE \leftarrow X)\ \$X)$$

- (BAG (STRIP x)) = x:

$$BAGSTRIP = (LAMBDA\ (BAG\ (STRIP \leftarrow X))\ \$X)$$

6.8. Reasoning about Substitutions

The rules in this section were added to prove assertions about the pattern matcher and the unification algorithm.

- An atom is either a variable or a constant:

$$\neg var(x) \wedge \neg const(x) \supset \neg atom(x):$$

$$NOTATOM = (LAMBDA\ (NOT\ (ATOM \leftarrow X))$$
$$(PROG\ (DECLARE)$$
$$(EXISTS\ (NOT\ (VAR\ \$X)))$$
$$(EXISTS\ (NOT\ (CONST\ \$X]$$

- If an expression is made of constants, so is the *car* and the *cdr* of the expression:

$$CONSTCAR = (LAMBDA\ (CONSTEXP\ (CAR \leftarrow X))$$
$$(EXISTS\ (CONSTEXP\ \$X]$$
$$CONSTCDR = (LAMBDA\ (CONSTEXP\ (CDR \leftarrow X))$$
$$(EXISTS\ (CONSTEXP\ \$X]$$

NOTATOM, CONSTCAR, and CONSTCDR are DEDUCE rules.

- The empty substitution does not change the expression:

$$SUBSTEMPTY = (LAMBDA\ (VARSUBST\ EMPTY \leftarrow X)\$X)$$

- No substitution changes an expression made up entirely of constants:

  SUBSTCONST=(LAMBDA (VARSUBST ←S ←Y)
              (PROG (DECLARE)
                  (GOAL $DEDUCE (CONSTEXP $Y))
                  $Y)

  SUBSTEMPTY and SUBSTCONST are simplification rules.

- To prove

$$varsubst(s, car(x)) = car(y),$$

prove

$$varsubst(s, x) = y:$$

  SUBSTCAR=(LAMBDA (EQ (VARSUBST ←S1 (CAR ←X))(CAR ←Y))
               (GOAL $EQRULES (EQ (VARSUBST $S1 $X)
                                    $Y]

- Similarly, to prove

$$varsubst(s, cdr(x)) = cdr(y),$$

prove

$$varsubst(s, x) = y:$$

  SUBSTCDR=(LAMBDA (EQ (VARSUBST ←S1 (CDR ←X))(CDR ←Y))
               (GOAL $EQRULES (EQ (VARSUBST $S1 $X)
                                    $Y]

- To prove

$$varsubst(s, x) = y,$$

where x and y are nonatomic, prove

$$varsubst(s, car(x)) = car(y)$$

and

$$varsubst(s, cdr(x)) = cdr(y):$$

  SUBSTCONS=
      (LAMBDA (EQ (VARSUBST ←S1 ←X) ←Y)
          (PROG (DECLARE)
              (GOAL $DEDUCE (NOT (ATOM $X)))
              (GOAL $DEDUCE (NOT (ATOM $Y)))

```
(GOAL (= ($REMOVE EQSUBST FROM $EQRULES))
      (EQ (VARSUBST $S1 (CAR $X))(CAR $Y)))
(GOAL (= ($REMOVE EQSUBST FROM $EQRULES))
     ( EQ (VARSUBST $S1 (CDR $X))(CDR $Y]
```

• SUBSTCAR, SUBSTCDR, and SUBSTCONS are equality rules. They are clustered together in a goal class:

EQSUBSTRULES=
(TUPLE SUBSTCAR SUBSTCDR SUBSTCARCDR SUBSTCONS)

• EQSUBSTRULES is called from EQSUBST, an EQRULE.

EQSUBST=(LAMBDA (PAND ←Y (EQ (VARSUBST ←S ←X) ←Z))
                (GOAL $EQSUBSTRULES $Y]

Note that SUBSTCONS removes EQSUBST from the EQRULES. This avoids the loop that would occur if SUBSTCAR were applied after SUBSTCONS.

• To prove

$$varsubst(s, u) = varsubst(s, v),$$

where u and v are nonatomic, prove

$$varsubst(s, car(u)) = varsubst(s, car(v))$$

and

$$varsubst(s, cdr(u)) = varsubst(s, cdr(v)):$$

```
SUBSTCARCDR=(LAMBDA (EQ (VARSUBST ←S ←U)
                        (VARSUBST ←S ←V))
            (PROG (DECLARE)
                  (GOAL $DEDUCE (NOT (ATOM $U)))
                  (GOAL $DEDUCE (NOT (ATOM $V)))
                  (GOAL $EQRULES
                       (EQ (VARSUBST $S (CAR $U))
                           (VARSUBST $S (CAR $V))))
                  (GOAL $EQRULES
                       (EQ (VARSUBST $S (CDR $U))
                           (VARSBUST $S (CDR $V]
```

Substitutions are represented as lists of dotted pairs.

• If v is a variable,

$$varsubst(((v{\cdot}y)), v) = y:$$

```
SUBSTLIST=(LAMBDA (VARSUBST (LIST (CONS  ←V ←Y))←V)
          (PROG (DECLARE)
          (EXISTS (VAR $V)) $Y]
```

- The composition operator has the property

$$varsubst(compose(s1, s2), x) = varsubst(s1, varsubst(s2, x)):$$

```
SUBSTCOMPOSE=(LAMBDA (VARSUBST (COMPOSE ←S1 ←S2)←X)
             ($TRY $SUBSTRULES
                  ('(VARSUBST $S1 (VARSUBST $S2 $X]
```

- These simplification rules are members of the goal class

```
SUBSTRULES=
          (TUPLE SUBSTEMPTY SUBSTLIST SUBSTCOMPOSE SUBSTCONST)
```

which is called by SUBSTOP, a member of TOPRULES:

```
SUBSTOP=(LAMBDA (PAND (VARSUBST ←←X)←Y)
        (GOAL $SUBSTRULES $Y]
```

6.9. Utility Functions

- TRY is like a GOAL statement that will not fail if none of the goal class applies but instead returns its argument.

```
TRY=(LAMBDA (TUPLE ←GOALCLASS ←GOAL1)
    (ATTEMPT (GOAL $GOALCLASS $GOAL1)
    ELSE $GOAL1]
```

It evaluates (GOAL $GOALCLASS $GOAL1), but, if failure results, it returns GOAL1.

- TRYALL will try a goal class on an expression. If any member of the goal class applies, it will apply the same goal class to the resulting expression, and so on, until no rule applies. TRYALL returns the last expression it has derived, which may be the same as the first expression; TRYALL will not fail:

```
TRYALL=
(LAMBDA (TUPLE ←GOALCLASS1 ←GOAL1)
       (PROG (DECLARE)
           TOP (ATTEMPT (SETQ ←GOAL1 (GOAL $GOALCLASS1
                                          $GOAL1))
                       THEN
                       (GO TOP))
           (RETURN $GOAL1]
```

- TRYALLFAIL is like TRYALL, except it will fail if none of the goal class applies to the argument.

```
TRYALLFAIL=(LAMBDA (TUPLE ←GOALCLASS1 ←GOAL1)
              ($TRYALL $GOALCLASS1
                  (GOAL $GOALCLASS1 $GOAL1]
```

- TRYSUB applies a goal class to a specially designated subexpression of the given expression:

```
TRYSUB=(LAMBDA (TUPLE ←GOALCLASS ON ←SUB IN ←EXP)
            (SUBST $EXP
                  (TUPLE $SUB ($TRYALL $GOALCLASS $SUB]
```

- INSIST fails if its argument is FALSE.

```
INSIST=(LAMBDA ←X (IF $X ELSE (FAIL]
```

- REMOVE removes a designated item from a tuple.

```
REMOVE=(LAMBDA (TUPLE ←X FROM ←Y)
            ((QUOTE (LAMBDA (TUPLE ←←U $X ←←V)
                        (TUPLE $$U $$V )))
            $Y]
```

- ASK queries the user:

```
ASK=(LAMBDA ←X
        (IF (LISP ASK $X)
        ELSE
        (FAIL]
```

It types two expressions, the first a QA4 expression and the second an atom (e.g., PROVE? or SIMPLIFY?). If the user types YES, TRUE, OK, Y, or T, ASK returns TRUE; otherwise ASK fails. ASK uses a LISP function of the same name.

- SHORTEST computes the "smallest" element of a set, bag, or tuple:

```
SHORTEST=
(LAMBDA ←X
  (PROG (DECLARE BEST BESTCOUNT TEMPCOUNT)
        (SETQ ←BESTCOUNT 2000)
        (MAPC $X (QUOTE (LAMBDA ←Y
                    (IF (OR (LT (SETQ ←TEMPCOUNT
                                (LISP QA4COUNT $Y))
```

TABLE 1.1
Index of Functions and Goal Classes

Name	Page	Name	Page	Name	Page
ACCEX	55	FSUBTRACT1	47	PROOFLEIB	41
ACCH	55	FSUBTRACT2	47	PROOFSIMP	42
ANDSPLIT	38	GCDEQ	55	PROOFSWITCH	39
ARGSIMP	45	GTQLTQ	42	PROVE[a]	38
ASK	67	HASSIMP	45	RELCHECK	40
BACH	58	INEQIFTHENELSE	43	REMOVE	67
BAGAEMPTY	57	INEQLEIB	42	SETSIMP	45
BAGAII	57	INEQSTRIPBAG	43	SHORTEST	67
BAGALOWERPLUS	58	INEQSTRIPSTRIP	61	SIMPONE	44
BAGAMINUS	57	INEQSTRIPTRAN	61	SQRULE	53
BAGAOP	57	INEQTIMESDIVIDE	48	SUBNUM	54
BAGAPLUS	57	INEQUALITIES[a]	41	SUBPLUS	54
BAGARULES[a]	56	INSIST	67	SUBSTCAR	64
BAGEX	59	LEIBB	40	SUBSTCARCDR	65
BAGEX1	59	LEIBF	40	SUBSTCDR	64
BAGEXCOMPLICATED	60	LEIBS	40	SUBSTCOMPOSE	66
BAGSIMP	45	LEIBT	40	SUBSTCONS	64
BAGSTRIP	63	LTPLUS	47	SUBSTCONST	64
CANCEL	53	LTQMANY	42	SUBSTEMPTY	63
CONSDIFF	62	LTQPLUS	47	SUBSTLIST	65
CONSTCAR	63	MAXONE	56	SUBSTOP	66
CONSTCDR	63	MAXPLUS	56	SUBSTRULES[a]	66
DEDUCE[a]	44	MINUSMINUS	54	TIMESDIVIDEONE	53
DIFDIF	62	MINUSOP	53	TIMESEMPTY	52
DIFFCONS	63	MINUSPLUS	54	TIMESEXP	53
DIFFONE	63	MINUSZERO	54	TIMESONE	52
DIFFXX	62	NOTATOM	63	TIMESOP	52
DOWNRULES[a]	45	ORSPLIT	39	TIMESPLUS	52
EQINEQMONOTONE	48	ORSPLITMANY	39	TIMESRULES[a]	52
EQNUMB	46	PLUSCOMBINE	51	TIMESSINGLE	52
EQRULES[a]	40	PLUSDIFFERENCE	51	TIMESTIMES	53
EQSIMP	41	PLUSEMPTY	50	TIMESZERO	52
EQSUBST	65	PLUSMINUS	51	TOPRULES[a]	45
EQSUBSTRULES[a]	65	PLUSNUMBER	51	TRY	66
EQTIMESDIVIDE	47	PLUSOP	50	TRYALL	66
EXPEXP	54	PLUSPLUS	51	TRYALLFAIL	67
EXPZERO	54	PLUSRULES[a]	50	TRYSUB	67
FAILINTODOWNRULES	46	PLUSSINGLE	50	TUPSIMP	45
FIFTHENELSE	46	PLUSZERO	50		

[a] Goal class.

$BESTCOUNT)
                (EQUAL (STYPE $Y)
                        NUMBER))
        THEN
        (SETQ ←BEST $Y)
        (SETQ ←BESTCOUNT $TEMPCOUNT)))))
    $BEST]

The size of an expression is roughly the number of atoms in the expression. It is computed by a LISP function, QA4COUNT. Numbers are assumed to be "smaller" than identifiers.

Table 1.1 gives an index of functions and goal classes.

7. Appendix: Traces of Solutions

7.1. The Maximum of an Array (First Verification Condition)

A complete trace of a proof performed by our system is presented below. The verification condition to be proved is derived from the program to compute the maximal element of an array. Although a proof is contained above in the body of the text, following the trace tells us exactly what rules were applied in the proof. Furthermore, we can see exactly what false starts were made by the system and what user interaction was required to keep the program on the right track.

This particular verification condition was derived from the loop path of the program. The hypotheses are

1 (CONTEXT (1 0) 1 0)
2 (ASSERT (EQ MAX (ACCESS A LOC)) WRT $VERICON)
3 TRUE
4 (ASSERT (LTQ (STRIP (BAGA A 0 I)) MAX) WRT $VERICON)
5 TRUE
6 (ASSERT (LTQ 0 LOC) WRT $VERICON)
7 TRUE
8 (ASSERT (LTQ LOC I) WRT $VERICON)
9 TRUE
10 (ASSERT (LTQ I N) WRT $VERICON)
11 TRUE
12 (DENY (LT N (PLUS 1 I)) WRT $VERICON)
13 FALSE
14 (DENY (LT MAX (ACCESS A (PLUS 1 I))) WRT $VERICON)
15 FALSE

Since the hypotheses for the different verification conditions of a program may contradict each other, each proof is done in a separate context. The name of that context is VERICON. That is the meaning of the phrase "WRT $VERI-CON" which follows our assertions and goal. [For this proof, VERICON is ((1 0) 1 0).] Assertions made with respect to one VERICON will not affect problems solved with respect to another.

```
16 (GOAL $PROVE (LTQ (STRIP (BAGA A 0 (PLUS 1 I))) MAX) WRT
        $VERICON)
17   LAMBDA PROOFSWITCH (LTQ (STRIP (BAGA A 0 (PLUS 1 I))) MAX)
```

When a traced function is applied to an argument, the trace says

LAMBDA ⟨function name⟩ ⟨argument⟩.

Some of the utility functions are not traced.

```
18   (GOAL $INEQUALITIES ($F $X))
19     LAMBDA RELCHECK (LTQ (STRIP (BAGA A 0 (PLUS 1 I))) MAX)
20     LAMBDA PROOFSIMP (LTQ (STRIP (BAGA A 0 (PLUS 1 I))) MAX)
21       LAMBDA ARGSIMP (LTQ (STRIP (BAGA A 0 (PLUS 1 I))) MAX)
22         LAMBDA SIMPONE (TUPLE STRIP (BAGA A 0 (PLUS 1 I))) MAX)
   (TUPLE (STRIP (BAGA A 0 (PLUS 1 I))) MAX)
   SIMPLIFY?
   :Y
```

The system asked us whether we wanted it to simplify

(TUPLE (STRIP (BAGA A 0 (PLUS 1 I))) MAX).

We said yes.

```
23       (GOAL $TOPRULES $GOAL1)
24         LAMBDA HASSIMP (TUPLE (STRIP (BAGA A 0 (PLUS 1 I))) MAX)
25         (FAIL)
26         LAMBDA EQNUMB (TUPLE (STRIP (BAGA A 0 (PLUS 1 I))) MAX)
27         (FAIL)
28       (GOAL $DOWNRULES $GOAL1)
29         LAMBDA TUPSIMP (TUPLE (STRIP (BAGA A 0 (PLUS 1 I))) MAX)
30           LAMBDA SIMPONE (STRIP (BAGA A 0 (PLUS 1 I)))
   (STRIP (BAGA A 0 (PLUS 1 I)))
   SIMPLIFY?
   :Y
31         (GOAL $TOPRULES $GOAL1)
32           LAMBDA HASSIMP (STRIP (BAGA A 0 (PLUS 1 I)))
33           (FAIL)
34           LAMBDA EQNUMB (STRIP (BAGA A 0 (PLUS 1 I)))
```

```
35        (FAIL)
36        (GOAL $DOWNRULES (GOAL1)
37          LAMBDA ARGSIMP (STRIP (BAGA A 0 (PLUS 1 I)))
38            LAMBDA SIMPONE (BAGA A 0 (PLUS 1 I))
  (BAGA A 0 (PLUS 1 I))
  SIMPLIFY?
  :Y
```

We have given the system permission to simplify (BAGA A 0 (PLUS 1 I)).

```
39        (GOAL $TOPRULES $GOAL1)
40          LAMBDA HASSIMP (BAGA A 0 (PLUS 1 I))
41          (FAIL)
42          LAMBDA BAGAOP (BAGA A 0 (PLUS 1 I))
43          (GOAL $GOALCLASS1 $GOAL1)
44            LAMBDA BAGAPLUS (BAGA A 0 (PLUS 1 I))
45            (GOAL $DEDUCE (LTQ $I (PLUS 1 $J)))
```

The system tries to prove that 0≤(PLUS 1 I).

```
46              LAMBDA RELCHECK (LTQ 0 (PLUS 1 I))
47              LAMBDA LTQPLUS (LTQ 0 (PLUS 1 I))
48              (GOAL $DEDUCE (AND (LTQ $I $J) (LTQ 0 $K)))
49                LAMBDA RELCHECK (AND (LTQ 0 I) (LTQ 0 1))
```

It breaks down the goal to 0 ≤ I and 0 ≤ 1.

```
50                LAMBDA ANDSPLIT (AND (LTQ 0 I)(LTQ 0 1))
51                (GOAL $GOALCLASS $X)
52                  LAMBDA RELCHECK (LTQ 0 1)
53                  RELCHECK = TRUE
```

When a function returns a value, the trace says

$$\langle \text{function name} \rangle = \langle \text{value} \rangle.$$

In this case, the system knew that $0 < 1$ by performing the corresponding LISP evaluation.

```
54                (GOAL $GOALCLASS $Y)
55                  LAMBDA RELCHECK (LTQ 0 I)
56                  RELCHECK = TRUE
```

The 0 ≤ I follows from the hypotheses 6 and 8.

```
57                ANDSPLIT = TRUE
58                LTQPLUS = TRUE
59                BAGAPLUS = (BAG (ACCESS A (PLUS 1 I)) (STRIP
                  (BAGA A 0 I)))
```

The system has succeeded in simplifying

$$(BAGA\ A\ 0\ (PLUS\ 1\ I))$$

to

$$(BAG\ (ACCESS\ A\ (PLUS\ 1\ I))\ (STRIP\ (BAGA\ A\ 0\ I))).$$

60	(GOAL $GOALCLASS1 $GOAL1)
61	BAGAOP = (BAG (ACCESS A (PLUS 1 I)) (STRIP (BAGA A 0 I)))
62	SIMPONE = (BAG (ACCESS A (PLUS 1 I)) (STRIP (BAGA A 0 I)))
63	ARGSIMP = (STRIP (BAG (ACCESS A (PLUS 1 I)) (STRIP (BAGA A 0 I))))
64	(GOAL $GOALCLASS $GOAL1)
65	LAMBA HASSIMP (STRIP (BAG (ACCESS A (PLUS 1 I)) (STRIP (BAGA A 0 I))))
66	(FAIL)
67	LAMBDA EQNUMB (STRIP (BAG (ACCESS A (PLUS 1 I)) (STRIP (BAGA A 0 I))))
68	(FAIL)
69	SIMPONE = (STRIP (BAG (ACCESS A (PLUS 1I)) (STRIP (BAGA A 0 I))))
70	TUPSIMP = (TUPLE (STRIP (BAG (ACCESS A (PLUS 1 I)) (STRIP (BAGA A 0 I)))) MAX)
71	(GOAL $GOALCLASS $GOAL1)
72	LAMBDA HASSIMP (TUPLE (STRIP (BAG (ACCESS A (PLUS 1 I)) (STRIP (BAGA A 0 I)))) MAX)
73	(FAIL)
74	LAMBDA EQNUMB (TUPLE (STRIP (BAG (ACCESS A (PLUS 1 I)) (STRIP (BAGA A 0 I)))) MAX)
75	(FAIL)
76	SIMPONE = (TUPLE (STRIP (BAG (ACCESS A (PLUS 1 I)) (STRIP (BAGA A 0 I)))) MAX)
77	ARGSIMP = (LTQ (STRIP (BAG (ACCESS A (PLUS 1 I)) (STRIP (BAGA A 0 I)))) MAX)

The problem now is to prove

$$(STRIP\ (BAG\ (ACCESS\ A\ (PLUS\ 1\ I))\ (STRIP\ (BAGA\ A\ 0\ I)))) \leqslant MAX:$$

78	(GOAL $GOALCLASS1 $X)
79	LAMBDA RELCHECK (LTQ (STRIP (BAG (ACCESS A (PLUS 1 I)) (STRIP (BAGA A 0 I)))) MAX)
80	RELCHECK=TRUE

But since the system already knows

$$(ACCESS\ A\ (PLUS\ 1\ I)) \leqslant MAX \quad \text{from (14)},$$

and

$$(\text{STRIP (BAGA A 0 I)}) \leqslant \text{MAX} \quad \text{from (4)},$$

the proof is complete:

```
81    PROOFSIMP = TRUE
82    (ASSERT ($F $X))
83    PROOFSWITCH = (LTQ (STRIP (BAGA A 0 (PLUS 1 I))) MAX)
84    (LTQ (STRIP (BAGA A 0 (PLUS 1 I))) MAX)
```

7.2. The Maximum of an Array (Second Verification Condition)

The following is the trace of the proof for another verification condition for the program that computes the maximal element of an array. This verification condition is derived from the halt path of the program.

```
1   (CONTEXT (1 0) 1 0)
2   (ASSERT (EQ MAX (ACCESS A LOC)) WRT $VERICON)
3   TRUE
4   (ASSERT (LTQ (STRIP (BAGA A 0 I)) MAX) WRT $VERICON)
5   TRUE
6   (ASSERT (LTQ 0 LOC) WRT $VERICON)
7   TRUE
8   (ASSERT (LTQ LOC I) WRT $VERICON
9   TRUE
10  (ASSERT (LTQ I N) WRT $VERICON)
11  TRUE
12  (ASSERT (LT N (PLUS 1 I)) WRT $VERICON)
```

There is a demon that knows that in the integer domain,

$$x < y \supset x+1 \leqslant y.$$

This demon is responsible for the assertion

13 (ASSERT (LTQ (PLUS 1 $X) $Y) WRT $VERICON)

The system now knows $N+1 \leqslant I+1$. This assertion wakes up another demon, which asserts that $N \leqslant I$:

14 (ASSERT (LTQ $Y $Z) WRT $VERICON)
15 TRUE

Since $I \leqslant N$ has also been asserted (10), the mechanism for storing ordering relations silently tells the system that $I = N$.

The system proceeds with the proof:

16 (GOAL $PROVE (LTQ (STRIP (BAGA A 0 N)) (ACCESS A LOC)) WRT
                $VERICON)
17   LAMBDA PROOFSWITCH (LTQ (STRIP (BAGA A 0 N)) (ACCESS A
                LOC))
18 (GOAL $INEQUALITIES ($F $X))
19   LAMBDA RELCHECK (LTQ (STRIP) BAGA A 0 N)) (ACCESS A LOC))
20   LAMBDA PROOFSIMP (LTQ (STRIP (BAGA A 0 N))‚(ACCESS A LOC))
21   LAMBDA ARGSIMP (LTQ (STRIP (BAGA A 0 N)) (ACCESS A LOC))
22   LAMBDA SIMPONE (TUPLE (STRIP (BAGA A 0 N)) (ACCESS A
                LOC))
   (TUPLE (STRIP (BAGA A 0 N)) (ACCESS A LOC))
   SIMPLIFY?
   :N
23   (FAIL)
24   (FAIL)
25   LAMBDA PROOFLEIB (LTQ (STRIP (BAGA A 0 N)) (ACCESS A LOC))
26   (EXISTS ($F ←Y))

The system searches the data base for an assertion of the form (LTQ ←Y),
i.e., the gross form of the goal we are trying to prove. It finds one [assertion
(4)] and asks us if it should try to prove that the argument of the assertion
it has found is equal to the argument of the goal:

   (EQ (TUPLE (STRIP (BAGA A 0 N)) (ACCESS A LOC)) (TUPLE (STRIP
                (BAGA A 0 I)) MAX))
   PROVE?

We say yes, the proof proceeds.

   :Y
27   (GOAL $EQRULES (EQ $X $Y))
28   LAMBDA RELCHECK (EQ (TUPLE (STRIP (BAGA A 0 N)) (ACCESS
                A LOC)) (TUPLE (STRIP (BAGA A 0 I)) MAX))
29   RELCHECK = TRUE
30   PROOFLEIB = TRUE
31 (ASSERT ($F $X))
32 PROOFSWITCH = (LTQ (STRIP (BAGA A 0 N)) (ACCESS A LOC))
33 (LTQ (STRIP (BAGA A 0 N)) (ACCESS A LOC))

The two tuples were found to be equal because N = I (from 10 and 14), and
MAX = A[LOC] . The proof is complete.

7.3. The Wensley Division Algorithm

The following is the complete trace of the proof included in the body of the
text:

1 (CONTEXT (1 0) 1 0)

2 (ASSERT (EQ AA (TIMES QQ YY)) WRT $VERICON)
3 TRUE
4 (ASSERT (EQ (TIMES 2 BB) (TIMES QQ DD)) WRT $VERICON)
5 TRUE
6 (ASSERT (LT PP (PLUS (TIMES QQ YY) (TIMES QQ DD))) WRT
              $VERICON)
7 TRUE
8 (ASSERT (LTQ (TIMES QQ YY) PP) WRT $VERICON)
9 TRUE
10 (ASSERT (LT PP (PLUS AA BB)) WRT $VERICON)
11 TRUE
12 (DENY (LT (DIVIDES DD 2) EE) WRT $VERICON)
13 FALSE

The goal is to prove PP < QQ*YY + QQ*(DD/2):

14 (GOAL $PROVE (LT PP (PLUS (TIMES QQ YY) (TIMES QQ (DIVIDES
              DD 2)))) WRT $VERICON
15 LAMBDA PROOFSWITCH (LT PP (PLUS (TIMES QQ YY) (TIMES QQ
              (DIVIDES DD 2))))
16 (GOAL $INEQUALITIES ($F $X))
17 LAMBDA RELCHECK (LT PP (PLUS (TIMES QQ YY) (TIMES QQ
              (DIVIDES DD 2))))
18   LAMBDA PROOFSIMP (LT PP (PLUS (TIMES QQ YY) (TIMES QQ
              (DIVIDES DD 2))))
19   LAMBDA ARGSIMP (LT PP (PLUS (TIMES QQ YY) (TIMES QQ
              (DIVIDES DD 2))))
20   LAMBDA SIMPONE (TUPLE PP (PLUS (TIMES QQ YY) (TIMES
              QQ (DIVIDES DD 2))))
21   LAMBDA ASK (TUPLE PP (PLUS (TIMES QQ YY) (TIMES QQ
              (DIVIDES DD 2)))) SIMPLIFY?)
   (TUPLE PP (PLUS (TIMES QQ YY) (TIMES QQ (DIVIDES DD 2))))
   SIMPLIFY?
   :NO
22   (FAIL)
23   (FAIL)
24   LAMBDA PROOFLEIB (LT PP (PLUS (TIMES QQ YY) (TIMES QQ
              (DIVIDES DD 2))))
25   (EXISTS ($F ←Y))
26   LAMBDA ASK (TUPLE (EQ (TUPLE PP (PLUS (TIMES QQ YY)
                     (TIMES QQ (DIVIDES DD 2))))
                     (TUPLE PP (PLUS (TIMES QQ YY)
                     (TIMES QQ DD)))) PROVE?)
   (EQ (TUPLE PP (PLUS (TIMES QQ YY) (TIMES QQ (DIVIDES DD 2))))
   (TUPLE PP (PLUS (TIMES QQ YY) (TIMES QQ DD))))
   PROVE?
   :NO

27   (FAIL)
28   LAMBDA ASK (TUPLE (EQ (TUPLE PP (PLUS (TIMES QQ YY)
                 (TIMES QQ (DIVIDES DD 2))))
                 (TUPLE PP (PLUS AA BB))) PROVE?)
  (EQ (TUPLE PP (PLUS (TIMES QQ YY) (TIMES QQ (DIVIDES DD 2))))
  (TUPLE PP (PLUS AA BB)))
  PROVE?
  :NO
29   (FAIL)
30   (LAMBDA INEQLEIB (LT PP (PLUS (TIMES QQ YY) (TIMES QQ
                 (DIVIDES DD 2))))
31   (EXISTS ($L (TUPLE ←LOWER ←UPPER)))
32   LAMBDA ASK (TUPLE PROVE (LTQ PP PP) AND (LTQ (PLUS
                 (TIMES QQ YY) (TIMES QQ DD))
                 (PLUS (TIMES QQ YY) (TIMES
                 QQ (DIVIDES DD 2))))?)
  PROVE
  (LTQ PP PP)
  AND
  (LTQ (PLUS (TIMES QQ YY) (TIMES QQ DD)) (PLUS (TIMES QQ YY)
  (TIMES QQ (DIVIDES DD 2))))
  ?
  :NO
33   (FAIL)

After several false starts, the system uses hypothesis (10), to generate two sub-goals: PP ≤ PP and AA + BB ≤ QQ*YY + QQ*(DD/2). We give our approval of this tactic:

34   LAMBDA ASK (TUPLE PROVE (LTQ PP PP) AND (LTQ (PLUS AA
                 BB)(PLUS (TIMES QQ YY) (TIMES QQ (DIVIDES DD 2)))) ?)
  PROVE
  (LTQ PP PP)
  AND
  (LTQ (PLUS AA BB) (PLUS (TIMES QQ YY) (TIMES QQ (DIVIDES DD
  2))))
  ?
  :YES
35   ASK = TRUE

It proves the first subgoal immediately.

36   (GOAL $INEQUALITIES (AND (LTQ $X $LOWER) (LTQ $UPPER $Y)))
37   LAMBDA ANDSPLIT (AND (LTQ PP PP) (LTQ (PLUS AA BB) (PLUS
                 TIMES QQ YY) (TIMES QQ (DIVIDES DD 2)))))
38   (GOAL $GOALCLASS $X)

39   LAMBDA RELCHECK (LTQ PP PP)
40   RELCHECK = TRUE
41   (GOAL $GOALCLASS (AND $$Y))
42   LAMBDA ANDSPLIT (AND (LTQ (PLUS AA BB)
                                  (PLUS (TIMES QQ YY)
                                  (TIMES QQ (DIVIDES DD 2)))))
43   (GOAL $GOALCLASS $X)
44   LAMBDA RELCHECK (LTQ (PLUS AA BB) (PLUS (TIMES QQ
                                  YY) (TIMES QQ (DIVIDES DD 2))))
45   LAMBDA INEQMONOTONE (LTQ (PLUS AA BB)(PLUS (TIMES
                                  QQ YY) (TIMES QQ (DIVIDES DD 2))))
46   LAMBDA ASK (TUPLE ((LTQ AA (TIMES QQ (DIVIDES DD
                                  2))) (LTQ BB (TIMES QQ YY))) PROVE?)
      ((LTQ AA (TIMES QQ DIVIDES DD 2))) (LTQ BB (TIMES QQ YY)))
      PROVE?
      :NO
47   (FAIL)
48   LAMBDA INEQMONOTONE (LTQ (PLUS AA BB) (PLUS (TIMES
                                  QQ YY) (TIMES QQ (DIVIDES DD 2))))
49   LAMBDA ASK (TUPLE ((LTQ AA (TIMES QQ YY)) (LTQ BB
                                  (TIMES QQ (DIVIDES DD 2)))) PROVE?)
      ((LTQ AA (TIMES QQ YY)) (LTQ BB (TIMES QQ (DIVIDES DD 2))))
      PROVE?

It divides the second subgoal into two sub-subgoals: $AA \leqslant QQ*YY$ and $BB \leqslant QQ*(DD/2)$:

      :YES
50   ASK = TRUE
51   (GOAL $GOALCLASS (AND ($F (TUPLE $W $Y)) ($F (TUPLE $X
                                  $Z))))
52   LAMBDA ANDSPLIT (AND (LTQ AA (TIMES QQ YY)) (LTQ BB
                                  (TIMES QQ (DIVIDES DD 2))))
53   (GOAL $GOALCLASS $X)
54   LAMBDA RELCHECK (LTQ AA (TIMES QQ YY))
55   RELCHECK = TRUE

The first sub-subgoal follows from hypothesis (2).

56   (GOAL $GOALCLASS (AND $$Y))
57   LAMBDA ANDSPLIT (AND LTQ BB (TIMES QQ (DIVIDES DD
                                  2))))
58   (GOAL $GOALCLASS $X)
59   LAMBDA RELCHECK (LTQ BB (TIMES QQ (DIVIDES DD
                                  2)))

```
60        LAMBDA INEQTIMESDIVIDE (LTQ BB (TIMES QQ (DIVIDES
               DD 2)))
61        (GOAL $DEDUCE (LT 0 $Y))
62         LAMBDA RELCHECK (LT 0 2)
63         RELCHECK = TRUE
64        (GOAL $INEQUALITIES ($F (TUPLE (TIMES $Y $W) (TIMES
               $X $$Z))))
65        LAMBDA RELCHECK (LTQ (TIMES 2 BB) (TIMES QQ DD))
66        RELCHECK = TRUE
```

Checking that $2 > 0$, the system multiplies out the second subgoal into $2*BB \leqslant QQ*DD$. This follows from assertion (4). The proof is complete:

```
67           INEQTIMESDIVIDE = TRUE
68          (GOAL $GOALCLASS (AND $$Y))
69           ANDSPLIT = (AND)
70          ANDSPLIT = (AND)
71         INEQMONOTONE = (AND)
72        (GOAL $GOALCLASS (AND $$Y))
73        ANDSPLIT = (AND)
74       ANDSPLIT = (AND)
75      INEQLEIB = (AND)
76     (ASSERT ($F $X))
77     (RETURN ($F $X))
78     PROOFSWITCH = (LT PP (PLUS (TIMES QQ YY) (TIMES QQ (DIVIDES
              DD 2))))
79     (LT PP (PLUS (TIMES QQ YY) (TIMES QQ (DIVIDES DD 2))))
```

7.4. The Pattern Matcher

As an abbreviation, let

$$ml = match(car(pat), car(arg))$$

and

$$m2 = match(varsubst(m1, cdr(pat)), cdr(arg)).$$

The hypotheses are that

$$varsubst(m1, car(pat)) = car(arg);$$

or, in unabbreviated form,

```
1  (ASSERT (EQ (VARSUBST (MATCH (CAR PAT) (CAR ARG))
   (CAR PAT)) (CAR ARG)))
2  TRUE
```

and that

$$varsubst(m2, varsubst(m1, cdr(pat))) = cdr(arg):$$

3 (ASSERT (EQ (VARSUBST (MATCH (VARSUBST (MATCH (CAR PAT)
   (CAR ARG)) (CDR PAT)) (CDR ARG)) (VARSUBST (MATCH (CAR PAT)
   (CAR ARG)) CDR PAT))) (CDR ARG)))
4 TRUE

The other hypotheses are

5   (ASSERT (CONSTEXP ARG))
6   TRUE
7   (ASSERT (NOT (CONST PAT)))
8   TRUE
9   (ASSERT (NOT (ATOM ARG)))
10 TRUE
11 (ASSERT (NOT (VAR PAT)))
12 TRUE

The goal is to prove

$$varsubst(compose(m2, m1), pat) = arg:$$

13 (GOAL $PROVE (EQ (VARSUBST (COMPOSE (MATCH (VARSUBST
   (MATCH (CAR PAT) (CAR ARG)) (CDR PAT)) (CDR ARG)) (MATCH
   (CAR PAT) (CAR ARG))) PAT) ARG))

The proof begins:

14 LAMBDA PROOFSWITCH (EQ (VARSUBST (COMPOSE (MATCH
   (VARSUBST (MATCH (CAR PAT) (CAR ARG)) (CDR PAT)) (CDR ARG))
   (MATCH (CAR PAT) (CAR ARG))) PAT) ARG)
15 (GOAL $EQRULES ($F $X))
16   LAMBDA RELCHECK (EQ (VARSUBST (COMPOSE (MATCH
   (VARSUBST (MATCH (CAR PAT) (CAR ARG)) (CDR PAT)) (CDR ARG))
   (MATCH (CAR PAT) (CAR ARG))) PAT) ARG)
17   LAMBDA EQSUBST (EQ (VARSUBST (COMPOSE (MATCH (VARSUBST
   (MATCH (CAR PAT) (CAR ARG)) (CDR PAT)) (CDR ARG)) (MATCH
   (CAR PAT) (CAR ARG))) PAT) ARG)
18   (GOAL $EQSUBSTRULES $Y)
19   LAMBDA SUBSTCONS (EQ (VARSUBST (COMPOSE (MATCH
   (VARSUBST (MATCH (CAR PAT) (CAR ARG)) (CDR PAT)) (CDR
   ARG)) (MATCH (CAR PAT) (CAR ARG))) PAT) ARG)
20   (GOAL $DEDUCE (NOT (ATOM $X)))
21   LAMBDA RELCHECK (NOT (ATOM PAT))

22   LAMBDA NOTATOM (NOT (ATOM PAT))
23   (EXISTS (NOT (VAR $X)))
24   (EXISTS (NOT (CONST $X)))
25   NOTATOM = (NOT (CONST PAT))
26   (GOAL $DEDUCE (NOT (ATOM $Y)))
27   (GOAL (= ($REMOVE (TUPLE EQSUBST FROM $EQRULES)))
        (EQ (VARSUBST $S1 (CAR $X)) (CAR $Y)))

Reasoning that *pat* is not an atom because it is neither a variable nor a constant, the system breaks the goal into two subgoals:

$$varsubst(compose(m2, m1), car(pat)) = car(arg)$$

and

$$varsubst(compose(m2, m1), cdr(pat)) = cdr(arg).$$

It begins work on the first of these:

28   LAMBDA RELCHECK (EQ (CAR ARG) (VARSUBST (COMPOSE
        (MATCH (VARSUBST (MATCH (CAR PAT) (CAR ARG)) (CDR
        PAT)) (CDR ARG)) (MATCH (CAR PAT) (CAR ARG))) (CAR PAT)))
29   LAMBDA EQSIMP (EQ (CAR ARG) (VARSUBST (COMPOSE
        (MATCH (VARSUBST (MATCH (CAR PAT) (CAR ARG)) (CDR PAT))
        (CDR ARG)) (MATCH (CAR PAT) (CAR ARG))) (CAR PAT)))
30   LAMBDA SIMPONE (VARSUBST (COMPOSE (MATCH (VARSUBST
        (MATCH (CAR PAT) (CAR ARG)) (CDR PAT)) (CDR ARG)) (MATCH
        (CAR PAT) (CAR ARG))) (CAR PAT))
   (VARSUBST (COMPOSE (MATCH (VARSUBST (MATCH (CAR PAT) (CAR
   ARG)) (CDR PAT)) (CDR ARG)) (MATCH (CAR PAT) (CAR ARG))) (CAR
   PAT))
   SIMPLIFY?

We give the system our permission to simplify the left side of the first subgoal,

$$varsubst(compose(m2, m1), car(pat)):$$

31   (GOAL $TOPRULES $GOAL1)
32   LAMBDA HASSIMP (VARSUBST (COMPOSE (MATCH (VARSUBST
        (MATCH (CAR PAT) (CAR ARG)) (CDR PAT)) (CDR ARG))
        (MATCH (CAR PAT) (CAR ARG))) (CAR PAT))
33   (FAIL)
34   LAMBDA SUBSTOP (VARSUBST (COMPOSE (MATCH
        (VARSUBST (MATCH (CAR PAT) (CAR ARG)) (CDR PAT))
        (CDR ARG)) (MATCH (CAR PAT) (CAR ARG))) (CAR PAT))
35   (GOAL $SUBSTRULES $Y)
36   LAMBDA SUBSTCOMPOSE (VARSUBST (COMPOSE (MATCH
        (VARSUBST (MATCH (CAR PAT) (CAR ARG)) (CDR PAT))
        (CDR ARG)) (MATCH (CAR PAT) (CAR ARG))) (CAR PAT))

37   (GOAL $GOALCLASS $GOAL1)
38   LAMBDA SUBSTCONST (VARSUBST (MATCH (VARSUBST
          (MATCH (CAR PAT) (CAR ARG)) (CDR PAT)) (CDR ARG))
          (VARSUBST (MATCH (CAR PAT) (CAR ARG)) (CAR PAT)))
39   (GOAL $DEDUCE (CONSTEXP $Y))
40   LAMBDA RELCHECK (CONSTEXP (VARSUBST (MATCH
          (CAR PAT) (CAR ARG)) (CAR PAT)))
41   SUBSTCOMPOSE = (VARSUBST (MATCH (VARSUBST (MATCH
          (CAR PAT) (CAR ARG)) (CDR PAT)) (CDR ARG)) (VARSUBST
          (MATCH (CAR PAT) (CAR ARG)) (CAR PAT)))
42   SUBSTOP = (VARSUBST (MATCH (VARSUBST (MATCH (CAR
          PAT) (CAR ARG)) (CDR PAT)) (CDR ARG)) (VARSUBST (MATCH
          (CAR PAT) (CAR ARG)) (CAR PAT)))
43   (RETURN $SIMPGOAL)
44   SIMPONE = (VARSUBST (MATCH (VARSUBST (MATCH (CAR PAT)
          (CAR ARG)) (CDR PAT)) (CDR ARG)) (VARSUBST (MATCH (CAR
          PAT) (CAR ARG)) (CAR PAT)))

The system has succeeded in simplifying the left half of the goal into

$$varsubst(m2, varsubst(m1, car(pat))).$$

It now tries to prove this new expression equal to $car(arg)$. In sympathy with our conscientious readers, we omit portions of the remainder of the trace.

55   LAMBDA SIMPONE (VARSUBST (MATCH (VARSUBST (MATCH
          (CAR PAT) (CAR ARG)) (CDR PAT)) (CDR ARG)) (VARSUBST
          (MATCH (CAR PAT) (CAR ARG)) (CAR PAT)))
          (VARSUBST (MATCH (VARSUBST (MATCH (CAR PAT) (CAR ARG))
          (CDR PAT)) (CDR ARG)) (VARSUBST (MATCH (CAR PAT) (CAR ARG))
          (CAR PAT)))
          SIMPLIFY?

The system asks permission to simplify

$$varsubst(m1, varsubst(m2, car(pat)))$$

further. Permission is granted. We omit a portion of the trace.

The system wants to simplify $varsubst(m1, car(pat))$, a subexpression of our goal. We give our blessings:

          :Y
80   (GOAL $TOPRULES $GOAL1)
81   LAMBDA HASSIMP·(VARSUBST (MATCH (CAR PAT)
          (CAR ARG)) (CAR PAT))
82   (FAIL)

83   LAMBDA SUBSTOP (VARSUBST (MATCH (CAR PAT)
                        (CAR ARG)) (CAR PAT))
84   (GOAL $SUBSTRULES $Y)
85   LAMBDA SUBSTCONST (VARSUBST (MATCH (CAR
                        PAT) (CAR ARG)) (CAR PAT))
86   (GOAL $DEDUCE (CONSTEXP $Y))
87   LAMBDA RELCHECK (CONSTEXP (CAR PAT))
88   LAMBDA CONSTCAR (CONSTEXP (CAR PAT))
89   (EXISTS (CONSTEXP $X))
90   LAMBDA EQNUMB (VARSUBST (MATCH (CAR
                        PAT) (CAR ARG)) (CAR PAT))
91   (RETURN $BEST)
92   EQNUMB = (CAR ARG)
93   (RETURN $SIMPGOAL)
94   SIMPONE = (CAR ARG)

The subexpression *varsubst*($m1$, *car*(*pat*)) is known to be equal to *car*(*arg*) by hypothesis (1). The rule EQNUMB has found this simplification. Work continues on simplifying the entire left-hand side.

95   TUPSIMP = (TUPLE (MATCH (VARSUBST (MATCH
                        (CAR PAT) (CAR ARG)) (CDR PAT)) (CDR ARG))
                        (CAR ARG))
96   GOAL $GOALCLASS $GOAL1)
97   LAMBDA HASSIMP (TUPLE (MATCH (VARSUBST
                        (MATCH (CAR PAT) (CAR ARG)) (CDR PAT)) (CDR
                        ARG)) (CAR ARG))
98   (FAIL)
99   LAMBDA EQNUMB (TUPLE (MATCH (VARSUBST
                        (MATCH (CAR PAT) (CAR ARG)) (CDR PAT)) (CDR
                        ARG)) (CAR ARG))
100   (FAIL)
101   (RETURN $SIMPGOAL)
102   SIMPONE = (TUPLE (MATCH (VARSUBST (MATCH
                        (CAR PAT) (CAR ARG)) (CDR PAT)) (CDR ARG))
                        (CAR ARG))
103   ARGSIMP = (VARSUBST (MATCH (VARSUBST (MATCH
                        (CAR PAT) (CAR ARG)) (CDR PAT)) (CDR ARG)) (CAR
                        ARG))

The expression being simplified is now *varsubst*($m2$, *car*(*arg*)):

104   (GOAL $GOALCLASS $GOAL1)
105   LAMBDA HASSIMP (VARSUBST (MATCH (VARSUBST
                        (MATCH (CAR PAT) (CAR ARG)) (CDR PAT)) (CDR
                        ARG)) (CAR ARG))

```
106              (FAIL)
107              LAMBDA SUBSTOP (VARSUBST (MATCH (VARSUBST
                 (MATCH (CAR PAT) (CAR ARG)) (CDR PAT)) (CDR
                 ARG)) (CAR ARG))
108              (GOAL $SUBSTRULES $Y)
109              LAMBDA SUBSTCONST (VARSUBST (MATCH
                 (VARSUBST (MATCH (CAR PAT) (CAR ARG)) (CDR
                 PAT)) (CDR ARG)) (CAR ARG))
110              (GOAL $DEDUCE (CONSTEXP $Y))
111              LAMBDA RELCHECK (CONSTEXP (CAR ARG))
112              LAMBDA CONSTCAR (CONSTEXP (CAR ARG))
113              (EXISTS (CONSTEXP $X))
114              CONSTCAR = (CONSTEXP ARG)
115              SUBSTCONST = (CAR ARG)
116              SUBSTOP = (CAR ARG)
117              (RETURN $SIMPGOAL)
118              SIMPONE = (CAR ARG)
```

Since *arg* consists entirely of constants, so does *car(arg)*. Therefore, substitutions have no effect on *car(arg)*, and the left-hand side of our subgoal reduces to *car(arg)* itself, which is precisely the same as the right-hand side.

```
119              (GOAL $EQRULES (EQ $X $Y))
120              LAMBDA RELCHECK (EQ (CAR ARG))
121              RELCHECK = TRUE
122              EQSIMP = TRUE
123              EQSIMP = TRUE
```

We omit the trace for the proof of the second subgoal,

$$varsubst(compose(m2, m1), cdr(pat)) = cdr(arg).$$

This subgoal is simplified to

$$varsubst(m2, varsubst(m1, cdr(pat))) = cdr(arg),$$

which is precisely our hypothesis (line 3).

```
142              (GOAL $EQRULES (EQ $X $Y))
143              EQSIMP = (EQ (VARSUBST (MATCH (VARSUBST (MATCH (CAR
                 PAT) (CAR ARG)) (CDR PAT)) (CDR ARG)) (VARSUBST (MATCH
                 (CAR PAT) (CAR ARG)) (CDR PAT))) (CDR ARG))
144              SUBSTCONS = (EQ (VARSUBST (MATCH (VARSUBST (MATCH
                 (CAR PAT) (CAR ARG)) (CDR PAT)) (CDR ARG)) (VARSUBST
                 (MATCH (CAR PAT) (CAR ARG)) (CDR PAT))) (CDR ARG))
145              EQSUBST = (EQ (VARSUBST (MATCH (VARSUBST (MATCH (CAR
                 PAT) (CAR ARG)) (CDR PAT)) (CDR ARG)) (VARSUBST (MATCH
                 (CAR PAT) (CAR ARG)) (CDR PAT))) (CDR ARG))
```

146   (ASSERT ($F $X))
147   (RETURN ($F $X))
148 PROOFSWITCH = (EQ (VARSUBST (COMPOSE (MATCH (VARSUBST
      (MATCH (CAR PAT) (CAR ARG)) (CDR PAT)) (CDR ARG)) (MATCH
      (CAR PAT) (CAR ARG))) PAT) ARG)
149 (EQ (VARSUBST (COMPOSE (MATCH (VARSUBST (MATCH (CAR PAT)
      (CAR ARG)) (CDR PAT)) (CDR ARG)) (MATCH (CAR PAT)
      (CAR ARG))) PAT) ARG)

The proof is complete.

7.5. The FIND Program

Only a selection from the trace for the interesting verification condition of
FIND is presented here because the entire trace was more than 500 lines long.
We will focus on the use of the case analysis during the proof.

The antecedent hypotheses for this condition are

$$1 \leqslant M \leqslant F \leqslant NN$$

$$M \leqslant I$$

$$J \leqslant N$$

$$(STRIP (BAGA A I M-1)) \leqslant (STRIP (BAGA A M NN))$$

$$(STRIP (BAGA A I N)) \leqslant (STRIP (BAGA A N+1 NN))$$

$$(STRIP (BAGA A 1 I-1)) \leqslant R$$

$$R \leqslant (STRIP (BAGA A 1+J NN))$$

$$A[J] \leqslant R$$

$$R \leqslant A[I]$$

$$I \leqslant J$$

$$J-1 < I+1$$

$$F \leqslant J-1$$

The theorem to be proved is

$$(STRIP (BAGA (EXCHANGE A I J) 1 J-1))$$

$$\leqslant (STRIP (BAGA (EXCHANGE A I J) (J-1)+1 \ NN).$$

This goal is simplified to

$$(IF \ J-1 < I \ THEN \ (STRIP (BAGA A 1 J-1))$$

$$ELSE \ (STRIP (BAG (STRIP (BAGA A 1 I-1))$$

$$A[J]$$

$$(STRIP \ (BAGA \ A \ I+1 \ J-1)))))$$

$$\leqslant (IF \ J \leqslant I \ THEN \ (STPIP) \ (BAGA \ A \ J \ NN)$$

$$ELSE \ (STRIP \ (BAG \ A[I]$$

$$(STRIP \ (BAGA \ A \ J+1 \ NN))$$

$$(STRIP \ (BAGA \ A \ J \ J-1)))))):$$

1 (GOAL $INEQUALITIES (LTQ (IFTHENELSE (LT (SUBTRACT J 1) I)
(STRIP (BAGA A 1 (SUBTRACT J 1))) (STRIP (BAG (STRIP (BAGA A 1
(SUBTRACT I 1))) (ACCESS A J) (STRIP (BAGA A (PLUS 1 I) (SUBTRACT
J 1)))))) (IFTHENELSE (LTQ J I) (STRIP (BAGA A J NN)) (STRIP (BAG
(ACCESS A I) (STRIP (BAGA A (PLUS 1 J) NN)) (STRIP (BAGA A J
(SUBTRACT J 1)))))))))

2 LAMBDA RELCHECK (LTQ (IFTHENELSE (LT (SUBTRACT J 1) I)
(STRIP (BAGA A 1 (SUBTRACT J 1))) (STRIP (BAG (STRIP (BAGA A 1
(SUBTRACT I 1))) (ACCESS A J) (STRIP (BAGA A (PLUS 1 I)
(SUBTRACT J 1)))))) (IFTHENELSE (LTQ J I) (STRIP (BAGA A J NN))
(STRIP (BAG (ACCESS A I) (STRIP (BAGA A (PLUS 1 J) NN)) (STRIP
(BAGA A J (SUBTRACT J 1)))))))

3 LAMBDA INEQIFTHENELSE (LTQ (IFTHENELSE (LT (SUBTRACT J 1)
I) (STRIP (BAGA A 1 (SUBTRACT J 1))) (STRIP (BAG (STRIP (BAGA
A 1 (SUBTRACT I 1))) (ACCESS A J) (STRIP (BAGA A (PLUS 1 I)
(SUBTRACT J 1)))))) (IFTHENELSE (LTQ J I) (STRIP (BAGA A J NN))
(STRIP (BAG (ACCESS A I) (STRIP (BAGA A (PLUS 1 J) NN)) (STRIP
(BAGA A J (SUBTRACT J 1)))))))

4 (ASSERT $X WRT $VERICON)

Since the left side of the goal has an IF-THEN-ELSE form, it causes the rule INEQIFTHENELSE to be applied. This rule sets VERICON to be a new lower context and asserts

$$J-1 < I$$

with respect to the new VERICON. This question triggers off a demon:

5 (ASSERT (LTQ (PLUS 1 $X) $Y) WRT $VERICON)

The new assertion is

$$(J-1)+1 \leqslant I.$$

The new assertion triggers off another demon, which makes still another assertion with respect to VERICON:

6 (ASSERT (LTQ $W $Y) WRT $VERICON)

This new assertion is

$$J \leqslant I.$$

(Later in the proof, another context will be established; $J-1 < I$ will be *denied* with respect to the new context.)

The THEN clause of the IF-THEN-ELSE expression must now be proved less than or equal to the right side of the goal:

7   (GOAL $INEQUALITIES ($F (TUPLE $$W1 $Y $$W2)) WRT $VERICON)

This goal is attempted with respect to the new context VERICON. In other words, we are trying to prove

(STRIP (BAGA A 1 J−1))

    $\leqslant$ (IF J $\leqslant$ I THEN (STRIP (BAGA A J NN))

        ELSE (STRIP (BAG A[I]

            (STRIP (BAGA A J+1 NN))

            (STRIP (BAGA A J J−1)))))

with respect to the context in which $J \leqslant I$ has been asserted:

8   LAMBDA RELCHECK (LTQ (STRIP (BAGA A 1 (SUBTRACT J 1)))
     (IFTHENELSE (LTQ J I) (STRIP (BAGA A J NN)) (STRIP (BAG
     (ACCESS A I) (STRIP (BAGA A (PLUS 1 J) NN)) (STRIP (BAGA
     A J (SUBTRACT J 1)))))))
9   LAMBDA INEQIFTHENELSE (LTQ (STRIP (BAGA A 1 (SUBTRACT J
     1))) (IFTHENELSE (LTQ J 1) (STRIP (BAGA A J NN)) (STRIP (BAG
     (ACCESS A I) (STRIP (BAGA A (PLUS 1 J) NN)) (STRIP (BAGA A J
     (SUBTRACT J 1)))))))

Since the right side of the inequality is still in IF-THEN-ELSE form, the rule INEQIFTHENELSE applies again. A new context VERICON, even lower than the last, is established, and the (redundant) statement

$$J \leqslant I$$

is asserted with respect to the new context:

10   (ASSERT $X WRT $VERICON)

A new goal is established with respect to the new context.

11   (GOAL $INEQUALITIES ($F (TUPLE $$W1 $Y $$W2)) WRT $VERICON)

The new goal is

$$\text{(STRIP (BAGA A 1 J}-1)) \leqslant \text{(STRIP (BAGA A J NN))}$$

12   LAMBDA RELCHECK (LTQ (STRIP (BAGA A 1 (SUBTRACT J 1)))
     (STRIP (BAGA A J NN)))
13   LAMBDA PROOFSIMP (LTQ (STRIP (BAGA A 1 (SUBTRACT J 1)))
     (STRIP (BAGA A J NN)))
14   LAMBDA ARGSIMP (LTQ (STRIP (BAGA A 1 (SUBTRACT J 1)))
     (STRIP (BAGA A J NN)))
15   LAMBDA SIMPONE (TUPLE (STRIP (BAGA A 1 (SUBTRACT J
     1))) (STRIP (BAGA A J NN)))

The simplifier is invoked. We will omit some steps from the trace here and mention only that the rule BAGALOWERPLUS played an important part in the simplification of the second element of the tuple.

90   SIMPONE = (TUPLE (STRIP (BAGA A 1 (SUBTRACT J 1))) (STRIP
       (BAG (STRIP (BAGA A (PLUS 1 J)) NN)) (ACCESS A J))))
91   ARGSIMP = (LTQ (STRIP (BAGA A 1 (SUBTRACT J 1))) (STRIP
       (BAG (STRIP (BAGA A (PLUS 1 J)) NN)) (ACCESS A J))))
92   (GOAL $GOALCLASS1 $X)

The simplified goal is

$$\text{(STRIP (BAGA A 1 J}-1))$$

$$\leqslant \text{(STRIP (BAG (STRIP (BAGA A J+1 NN)) A[J])):}$$

93   LAMBDA RELCHECK (LTQ (STRIP (BAGA A 1 (SUBTRACT J 1)))
     (STRIP (BAG (STRIP (BAGA A (PLUS 1 J) NN)) (ACCESS A J))))
94   LAMBDA INEQSTRIPBAG (LTQ (STRIP (BAGA A 1 (SUBTRACT
     J 1))) (STRIP (BAG (STRIP (BAGA A (PLUS 1 J) NN)) (ACCESS A
     J))))

INEQSTRIPBAG breaks up the goal into two subgoals. The first of these goals is

$$\text{(STRIP (BAGA A 1 J}-1)) \leqslant \text{(STRIP (BAGA A J+1 NN)):}$$

95   (GOAL $INEQUALITIES ($F (TUPLE $$W $X $$Z)))
96   LAMBDA RELCHECK (LTQ (STRIP (BAGA A 1 (SUBTRACT J 1)))
     (STRIP (BAGA A (PLUS 1 J) NN)))

The rule INEQSTRIPSTRIP is applicable to this goal:

103   LAMBDA INEQSTRIPSTRIP (LTQ (STRIP (BAGA A 1 (SUBTRACT
       J 1))) (STRIP (BAGA A (PLUS 1 J) NN)))

Since it is known that

$$(STRIP \ (BAGA \ A \ 1 \ I-1)) \leqslant (STRIP \ (BAGA \ A \ J+1 \ NN)),$$

and, in this context, $J \leqslant I$, INEQSTRIPSTRIP succeeds. The other subgoal to be proved is

$$(STRIP \ (BAGA \ A \ 1 \ J-1)) \leqslant (STRIP \ (BAG \ A[J])):$$

143   (GOAL $INEQUALITIES ($F (TUPLE $$W (STRIP (BAG $$Y))
$$Z)))

144   LAMBDA RELCHECK (LTQ (STRIP (BAGA A 1 (SUBTRACT
J 1))) (STRIP (BAG (ACCESS A J))))

INEQSTRIPBAG applies again, splitting this goal into two subgoals, one of which is trivial.

145   LAMBDA INEQSTRIPBAG (LTQ (STRIP (BAGA A 1
(SUBTRACT J 1))) (STRIP (BAG (ACCESS A J))))

146   (GOAL $INEQUALITIES ($F (TUPLE $$W $X $$Z)))

The nontrivial goal is

$$(STRIP \ (BAGA \ A \ 1 \ J-1)) \leqslant A[J]:$$

147   LAMBDA RELCHECK (LTQ (STRIP (BAGA A 1
(SUBTRACT J 1))) (ACCESS A J))

This goal invokes the rule INEQSTRIPTRAN. We will examine the operation of this rule in detail:

154   LAMBDA INEQSTRIPTRAN (LTQ (STRIP (BAGA A 1
(SUBTRACT J 1))) (ACCESS A J))

155   (EXISTS ($F (TUPLE (STRIP (BAGA $ARNAME ←LOWER
←UPPER)) ←D)))

The rule finds the hypothesis

$$(STRIP \ (BAGA \ A \ 1 \ I-1)) \leqslant R.$$

It tests if this relation is appropriate:

156   (GOAL $DEDUCE (AND (LTQ $LOWER $L) (LTQ $M
$UPPER) (LTQ $D $C)))

157   LAMBDA RELCHECK (AND (LTQ 1 1) (LTQ (SUBTRACT
J 1) (SUBTRACT I 1)) (LTQ R (ACCESS A J)))

The system is testing whether the array segment between 1 and $I-1$ includes the segment between 1 and J−1, and also whether $R \leqslant A[J]$:

158	LAMBDA ANDSPLIT (AND (LTQ 1 1) (LTQ (SUBTRACT J 1) (SUBTRACT I 1)) (LTQ R (ACCESS A J)))
159	(GOAL $GOALCLASS $X)
160	LAMBDA RELCHECK (LTQ 1 1)
161	RELCHECK = TRUE
162	(GOAL $GOALCLASS (AND $$Y))
163	LAMBDA RELCHECK (AND (LTQ (SUBTRACT J 1) (SUBTRACT I 1)) (LTQ R (ACCESS A J)))
164	LAMBDA ANDSPLIT (AND (LTQ (SUBTRACT J 1) (SUBTRACT I 1)) (LTQ R (ACCESS A J)))
165	(GOAL $GOALCLASS $X)
166	LAMBDA RELCHECK (LTQ (SUBTRACT J 1) (SUBTRACT I 1))
167	RELCHECK = TRUE
168	(GOAL $GOALCLASS (AND $$Y))
169	LAMBDA RELCHECK (AND (LTQ R (ACCESS A J)))
170	LAMBDA ANDSPLIT (AND (LTQ R (ACCESS A J)))
171	(GOAL $GOALCLASS $X)
172	LAMBDA RELCHECK (LTQ R (ACCESS A J))
173	RELCHECK = TRUE
174	(GOAL $GOALCLASS (AND $$Y))
175	ANDSPLIT = (AND)
176	ANDSPLIT = (AND)
177	ANDSPLIT = (AND)

The tests prove to be successful, and **INEQSTRIPTRAN** returns

178	INEQSTRIPTRAN = (AND)

The trivial subgoal is achieved:

179	(GOAL $INEQUALITIES ($F (TUPLE $$W (STRIP (BAG $$Y)) $$Z)))
180	LAMBDA RELCHECK (LTQ (STRIP (BAGA A 1 (SUBTRACT J 1))) (STRIP (BAG)))
181	RELCHECK = TRUE

The call to **INEQSTRIPBAG** from line 145 returns successfully:

182	INEQSTRIPBAG = TRUE

The call to **INEQSTRIPBAG** from line 94 also returns:

183	INEQSTRIPBAG = TRUE

Thus, the goal established in line 11 has been successfully proved:

184   PROOFSIMP = TRUE

That goal was established by the rule INEQIFTHENELSE. This rule asserted
J ⩽ I with respect to a lower context and set up the goal with respect to that
context. The rule now attempts to deny J ⩽ I with respect to another context
and to establish a new goal with respect to the new context.

185   (DENY $X WRT $VERICON)

However, J ⩽ I was also asserted with respect to a higher context in line 6.
Therefore, denying J ⩽ I contradicts this assertion, causing the denial to fail.
Since the situation is contradictory and could not arise, it is unnecessary to
achieve the goal, and the call to INEQIFTHENELSE from line 9 returns
successfully:

186   (RETURN (SUCCESS (TUPLE WITH INEQIFTHENELSE)))
187   INEQIFTHENELSE = (SUCCESS (TUPLE WITH INEQIFTHENELSE))

The goal established in line 7 has been achieved. This goal was set up by an
earlier call to INEQIFTHENELSE (line 3) with respect to a context in which
J−1 < I was asserted (line 4). It is now necessary to set up a new goal with
respect to a new context; in this new context, J−1 < I is denied:

188   (DENY $X WRT $VERICON)

This denial activates a demon that denies

$$J{\leqslant}I:$$

189   LAMBDA TRYALL (TUPLE (TUPLE PLUSEMPTY PLUSSINGLE
       PLUSZERO PLUSPLUS PLUSMINUS PLUSDIFFERENCE
       PLUSCOMBINE PLUSNUMBER) (PLUS 1 I (MINUS 1)))
190   (GOAL $GOALCLASS1 $GOAL1)
191   LAMBDA PLUSMINUS (PLUS 1 I (MINUS 1))
192   PLUSMINUS = (PLUS 1)
193   (GOAL $GOALCLASS1 $GOAL1)
194   LAMBDA PLUSSINGLE (PLUS I)
195   PLUSSINGLE = I
196   (GOAL $GOALCLASS1 $GOAL1)
197   (RETURN $GOAL1)
198   TRYALL = I
199 (DENY (LTQ $W $RTSIDE) WRT $VERICON)

The new goal

$$(\text{STRIP (BAG (STRIP (BAGA A 1 I−1)}$$
$$A[J]$$
$$(\text{STRIP (BAGA A I+1 J−1))))}$$
$$\leqslant (\text{IF J} \leqslant \text{I THEN (STRIP (BAGA A J NN))}$$
$$\text{ELSE (STRIP (BAG A[I]}$$
$$(\text{STRIP (BAGA A J+1 NN))}$$
$$(\text{STRIP (BAGA A J J−1))))}$$

is established with respect to the new context:

200 (GOAL $INEQUALITIES ($F (TUPLE $$W1 $Z $$W2)) WRT $VERICON)
201   LAMBDA RELCHECK (LTQ (STRIP (BAG (STRIP (BAGA A 1
(SUBTRACT I 1))) (ACCESS A J) (STRIP (BAGA A (PLUS 1 I)
(SUBTRACT J 1))))) (IFTHENELSE (LTQ J I) (STRIP (BAGA A J NN))
(STRIP (BAG (ACCESS A I) (STRIP (BAGA A (PLUS 1 J) NN)) (STRIP
(BAGA A J (SUBTRACT J 1)))))))

INEQIFTHENELSE is invoked because the right-side of the goal is of the form
IF-THEN-ELSE.

202   LAMBDA INEQIFTHENELSE (LTQ (STRIP (BAG (STRIP (BAGA A 1
(SUBTRACT I 1))) (ACCESS A J) (STRIP (BAGA A (PLUS 1 I)
(SUBTRACT J 1))))) (IFTHENELSE (LTQ J I) (STRIP (BAGA A J NN))
(STRIP (BAG (ACCESS A I) (STRIP (BAGA A (PLUS 1 J) NN)) (STRIP
(BAGA A J (SUBTRACT J 1)))))))

Again the rule creates two contexts: In one context J ≤ I is asserted, and in the
other J ≤ I is denied. However, since J ≤ I was denied in a higher context (line
199), the assertion of J ≤ I fails; this contradictory case can safely be ignored,
and attention focuses on the second context:

204   (DENY $X WRT $VERICON)

The goal is established using the ELSE clause of the previous goal,

$$(\text{STRIP (BAG (STRIP (BAGA A 1 I−1))}$$
$$A[J]$$
$$(\text{STRIP (BAGA A I+1 J−1))))}$$
$$\leqslant (\text{STRIP (BAG A[I]}$$

(STRIP (BAGA A J+1 NN))

(STRIP (BAGA A J J−1))))):

205 (GOAL $INEQUALITIES ($F (TUPLE $$W1 $Z $$W2)) WRT
$VERICON)
206   LAMBDA RELCHECK (LTQ (STRIP (BAG (STRIP (BAGA A 1
(SUBTRACT I 1))) (ACCESS A J) (STRIP (BAGA A (PLUS 1 I)
(SUBTRACT J 1))))) (STRIP (BAG (ACCESS A I) (STRIP (BAGA A
(PLUS 1 J) NN)) (STRIP (BAGA A J (SUBTRACT J 1))))))
207   LAMBDA INEQSTRIPBAG (LTQ (STRIP (BAG (STRIP (BAGA A 1
(SUBTRACT I 1))) (ACCESS A J) (STRIP (BAGA A (PLUS 1 I)
(SUBTRACT J 1))))) (STRIP (BAG (ACCESS AI) (STRIP (BAGA A
(PLUS 1 J) NN)) (STRIP (BAGA A J (SUBTRACT J 1))))))

The proof from this point will only be summarized, since it is lengthy but uneventful. The goal is divided into nine subgoals by successive applications of INEQSTRIPBAG. Each of these goals turns out to be easily proved, and the proof ends successfully.

558   INEQIFTHENELSE = TRUE
559 INEQIFTHENELSE = TRUE
560
TRUE

Acknowledgments

The work on program verification was done in close collaboration with Bernie Elspas. The work on QA4 was done with Jeff Rulifson and Jan Derksen. Irene Greif wrote the first version of the simplifier and participated in the conceptualization of the pattern matcher and unification proofs. Jeff Rulifson encouraged us to write this paper and suggested its format. Rich Fikes has helped with design modification and debugging of QA4. Bert Raphael read the manuscript and suggested many improvements. Tony Hoare read and commented on the entire final draft; much of his advice was incorporated in the published paper. This work has benefitted from our conversations with Cordell Green, Peter Neumann, Larry Robinson, Earl Sacerdoti, René Reboh, Mark Stickel, Steve Crocker, and John McCarthy. Many members of the Artificial Intelligence Center and the Computer Science Group at SRI helped with support and criticism.

The research reported herein was supported in part by the National Science Foundation under Grant GJ-36146 and in part by the Advanced Research Projects Agency under Contract DAHC04-72-C-0008 and by the National Aeronautics Space Administration under Contract NASW-2086.

Chapter 2
Logical Analysis of Programs

Shmuel Katz and Zohar Manna

1. Introduction

In recent years considerable effort has been devoted to the goal of proving (or "verifying") that a given computer program is partially correct—i.e., that if the program terminates, it satisfies some user-provided *input/output specification.* Floyd (1967) suggested a method for proving partial correctness of flowchart programs which has been shown amenable to mechanization [e.g., see King (1969), Deutsch (1973), Waldinger and Levitt (1974), Good et al. (1975), and Suzuki (1975)]. However, most existing implementations are incomplete in that they are not oriented toward incorrect programs: their declared goal is to prove that a correct program really is correct. If a program is not verified, it is unclear whether the program is erroneous or whether a proper proof has simply not been discovered.

We suggest conducting logical analysis of programs using "invariants" which express the actual relationships among the variables of the program. These "invariants" differ from Floyd's programmer-supplied "assertions" in that they are generated directly from the program text. In our conception, the invariants are independent of the output specification of the program and reflect what is actually happening during the computation, as opposed to what is supposed to be happening. Thus our invariants can be used either to verify the program with respect to its specifications or to prove that the program cannot be verified (i.e., contains an error). In addition, these

This is a revised version of a previously published article by the same name which appeared in *Communications of the ACM,* vol. 19, no. 4, pp. 188-206. Copyright 1976 by Association for Computing Machinery, Inc., reprinted by permission.

invariants enable us to integrate proofs of termination and non-termination into our logical analysis. Invariants can also be used to debug an incorrect program, i.e., to diagnose the errors and to modify the program.

The need to relieve the user of the task of supplying fully detailed assertions (or invariants) has been widely recognized. We devote a large part of the paper to presenting techniques for the systematic generation of invariants.

Ultimately, we envision a system based on these techniques which would automatically generate the straightforward invariants. The programmer would still be expected to supply suggestions for those invariants requiring more insight into the logic of the program. Whenever new invariants had been produced, all invariants generated up to that point would be used to check simultaneously for correctness or incorrectness. If correctness (including termination) has been established, an attempt may be made to optimize the program through a fundamental revision of the program statements, based on the invariants. If incorrectness has been established, an attempt is made to automatically debug the program, i.e., to diagnose and correct the errors in a systematic manner, again using the invariants. If neither correctness nor incorrectness can be established, we attempt to generate additional invariants and repeat the process. Assertions ("comments") supplied by the programmer may or may not be correct, and therefore are considered to be just promising candidate invariants. As a last resort, it may nevertheless be possible to take a more radical approach and use the invariants for modifying the program so that correctness is guaranteed, taking the calculated risk of modifying an already correct program.

In the following sections we first present the techniques of automatic invariant generation, an algorithmic approach in Section 3, and a heuristic approach in Section 4. Then in Section 5 we describe the applications of the invariants for proving correctness (including termination) or incorrectness. In Section 6 we outline the practical implications of the invariants for automatic debugging. The Conclusion includes some bibliographical remarks.

2. Preliminaries

The programs treated in this paper are written in a simple flow-chart language with standard arithmetic operators over the domain of real or integer numbers. We assume a flowchart program P with input variables $\overline{x}$, which do not change during execution, and program variables $\overline{y}$, which do change during execution and whose final values constitute the output of the program. In addition, we are given an *input predicate* $\phi(\overline{x})$, which restricts the legal input values, and an *output predicate* $\psi(\overline{x}, \overline{y})$, which indicates the desired relationship between the input and output values.

For convenience we consider *blocked programs*. That is, we assume the program is divisible into (possibly nested) "blocks" in such a way that every block has at most one top-level loop (in addition to possible lower-level loops which are already contained in inner blocks). The blocks we consider have one entrance and may have many exists. Every "structured program" can be decomposed into such blocks.

The block structure allows us to treat the program by first considering inner blocks (ignoring momentarily that they are included in outer blocks) and then working outwards. Thus, for each block we can consider its top-level loop using information we have obtained from the inner blocks.

The top-level loop of a block can contain several branches, but all paths around the loop must have at least one common point. For each loop we will choose one such point as the *cutpoint* of the loop.

We use *counters* attached to each block containing a loop as an essential tool in our techniques. Since each loop has a unique cutpoint, we associate a counter with the cutpoint of the loop. The counter is initialized before entering the block so that its value is 0 upon first reaching the cutpoint, and is incremented by 1 exactly once somewhere along the loop before returning to the cutpoint. There are many locations where the initialization of the counters could be done. The two extreme cases are of special interest: (1) the counter is initialized only once, at the beginning of the program (a "global" initialization, parametrizing the total number of times the cutpoint is reached), or (2) the counter is initialized just before entering its block (a "local" initialization, indicating the number of executions of the corresponding loop since the most recent entrance to the block). In the continuation, we will assume a *local* initialization of counters, since our experience has been that this is generally the most convenient choice.

The counters will play a crucial role both for generating invariants

and for proving termination. They will be used both to denote relations among the number of times various paths have been executed
and to help express the values assumed by the program variables. It
should be noted that it is unnecessary to add the counters physically
to the body of the program. Their location can merely be indicated,
since their behavior is already fixed.

It is sometimes convenient to add auxiliary cutpoints at the entrance and exit of a block. In addition, we always add a special cutpoint on each arc immediately preceding a HALT statement. Such
cutpoints will be called *haltpoints* of the program.

A typical situation is shown in Figure 2.1. There is an inner block
with cutpoint M and counter m, and an outer block with a cutpoint
N and counter n. The outer block also has auxiliary cutpoints L and
K at the entrance and exit of the block, respectively.

Our first task is to attach an appropriate invariant $q_i(\overline{x}, \overline{y})$ to each
cutpoint i. We first define our terms.

A predicate $q_i(\overline{x}, \overline{y})$ is said to be an *invariant at cutpoint i w.r.t.*
$\phi(\overline{x})$ if for every input $\overline{a}$ such that $\phi(\overline{a})$ is true, whenever we reach
point i with $\overline{y} = \overline{b}$, then $q_i(\overline{a}, \overline{b})$ is true. An invariant at i is thus
some statement about the variables which is true for the current
values of the variables each time i is reached during execution.

For a path α from cutpoint i to cutpoint j, we define $R_\alpha(\overline{x}, \overline{y})$
as the condition for the path α to be traversed, and $r_\alpha(\overline{x}, \overline{y})$ as
the transformation in the $\overline{y}$ values which occurs on path α. A set
S of points of a program P is said to be *complete* if, for each cutpoint i in S, all the cutpoints on any path from START to i are
also in S. For example, if L is the entrance to the program in Figure
2.1, $\{L\}$, $\{L, M, N\}$ and $\{L, M, N, K\}$ are all complete sets of cutpoints; $\{L, N\}$ is not.

We shall use the following sufficient condition [proven in Manna
(1969)] for showing that assertions ("candidate invariants") are
actually invariants.

Lemma A.   *Let S be a complete set of cutpoints of a program P.
Assertions $\{q_i(\overline{x}, \overline{y}) \mid i \in S\}$ will be a set of invariants for P w.r.t.*
ϕ *if*

(a) *for every path α from the* START *statement to a cutpoint*

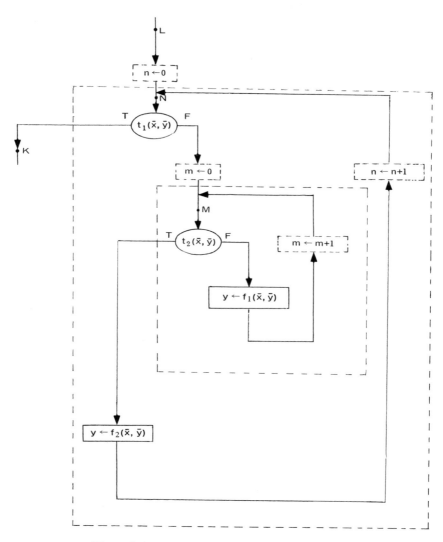

Figure 2.1. Blocks, Cutpoints, and Counters.

j (which does not contain any other cutpoint) [1] and

$$\forall \overline{x}[\phi(\overline{x}) \wedge R_\alpha(\overline{x}) \supset q_j(\overline{x}, r_\alpha(\overline{x}))],$$

(b) *for every path α from a cutpoint i to a cutpoint j (which does not contain any other cutpoint),*

$$\forall \overline{x} \forall \overline{y}[q_i(\overline{x}, \overline{y}) \wedge R_\alpha(\overline{x}, \overline{y}) \supset q_j(\overline{x}, r_\alpha(\overline{x}, \overline{y}))].$$

Consider, for example, the initial segment of a program shown in Figure 2.2. Assertions $q_1(\overline{x}, \overline{y})$ and $q_2(\overline{x}, \overline{y})$ will be invariants at cutpoints 1 and 2, respectively, if

(a) $\forall \overline{x}[\phi(\overline{x}) \supset q_1(\overline{x}, g(\overline{x}))]$,
(b) $\forall \overline{x} \forall \overline{y}[q_1(\overline{x}, \overline{y}) \wedge \sim t(\overline{x}, \overline{y}) \supset q_1(\overline{x}, f(\overline{x}, \overline{y}))]$,
   $\forall \overline{x} \forall \overline{y}[q_1(\overline{x}, \overline{y}) \wedge \ \ t(\overline{x}, \overline{y}) \supset q_2(\overline{x}, h(\overline{x}, \overline{y}))]$.

Intuitively, condition (a) implies that $q_1(\overline{x}, \overline{y})$ holds the first time the cutpoint 1 is reached, while the first formula of (b) states that if $q_1(\overline{x}, \overline{y})$ holds at the cutpoint, it is still true after the loop is executed. Thus, by induction it follows that $q_1(\overline{x}, \overline{y})$ holds whenever cutpoint 1 is reached, i.e., it is an invariant at 1. The second formula of (b) states that if $q_1(\overline{x}, \overline{y})$ holds at cutpoint 1, then $q_2(\overline{x}, \overline{y})$ holds at cutpoint 2. Thus, since $q_1(\overline{x}, \overline{y})$ is an invariant at 1, $q_2(\overline{x}, \overline{y})$ is an invariant at 2.

Note that the input predicate $\phi(\overline{x})$, which depends only on $\overline{x}$ (variables not changed during execution), is automatically an invariant of any cutpoint of the program, and does not need any further justification.

Lemma A is slightly misleading, because it implies that a fullfledged set of assertions is provided at a complete set of the cutpoints and that these are checked simultaneously. In practice, the invariants will be added one after another until the needs of the logical analysis have been met. At every stage of the invariant generating process, a situation as in Figure 2.3 will apply for each block. At cutpoints L, N, and M, invariants $p(\overline{x}, \overline{y})$, $q(\overline{x}, \overline{y})$, and $s(\overline{x}, \overline{y})$, respectively will already have been proven. However, we also will

---

[1] Note that the $\overline{y}$ values are not defined at the START statement, and that they are initialized by constants or functions of $\overline{x}$ along a path from START. Thus, R_α and r_α for such a path are really only functions of $\overline{x}$, and not of $\overline{y}$.

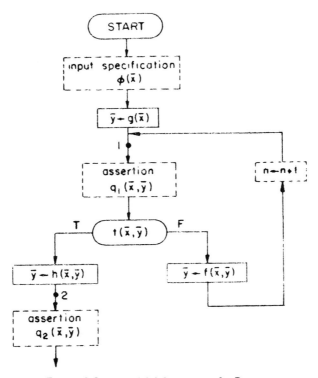

Figure 2.2. An Initial Segment of a Program.

have promising candidate invariants $p'(\bar{x}, \bar{y})$, $q'(\bar{x}, \bar{y})$ and $s'(\bar{x}, \bar{y})$, which we have so far been unable to prove invariant. These candidates could originate as comments given by the user or, as in the case of $s'(\bar{x}, \bar{y})$, from the output specification, which we automatically designate as a candidate at the haltpoints. As indicated in Section 4, additional candidates may be generated during this process.

For a block of the form given in Figure 2.3, we concentrate on developing invariants at cutpoint N on the loop. For the auxiliary cutpoint M, the invariants are generated by "pushing forward" any invariant obtained at N. Thus, if at any stage an invariant $q(\bar{x}, \bar{y})$ has been established at N, we automatically can take as an invariant at M any $s(\bar{x}, \bar{y})$ satisfying

$$\forall \bar{x} \, \forall \bar{y} [q(\bar{x}, \bar{y}) \wedge t(\bar{x}, \bar{y}) \supset s(\bar{x}, h(\bar{x}, \bar{y}))].$$

To establish that a candidate $q'(\bar{x}, \bar{y})$ is actually an invariant at N,

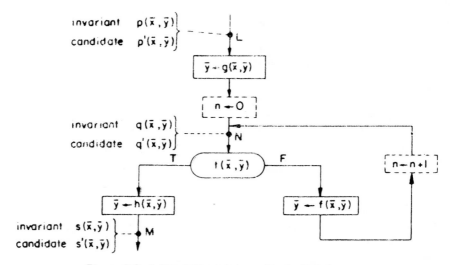

Figure 2.3. A Block Containing a Single-Path Loop.

it follows from Lemma A that we must show

(i) $\forall \bar{x} \, \forall \bar{y} [p(\bar{x}, \bar{y}) \supset q'(\bar{x}, g(\bar{x}, \bar{y}))]$

and

(ii) $\forall \bar{x} \, \forall \bar{y} [q(\bar{x}, \bar{y}) \wedge q'(\bar{x}, \bar{y}) \wedge \sim t(\bar{x}, \bar{y}) \supset q'(\bar{x}, f(\bar{x}, \bar{y}))]$.

It must be emphasized that special care should be taken in case of failure in an attempt to establish that a candidate is an invariant. For example, suppose that q'_1 and q'_2 are candidate invariants at the cutpoint N and that both q'_1 and q'_2 satisfy condition (i). It is entirely possible that neither q'_1 nor q'_2 satisfies condition (ii) individually, but that $q'_1 \wedge q'_2$ does satisfy condition (ii), and therefore is an invariant. This phenomenon—i.e., that it is impossible to show a weak property but it is possible to show a stronger one—is typical in mathematical proofs by induction. The explanation is that although we must show a stronger property on the right of the implication, we are also provided with a stronger inductive hypothesis on the left.

In Sections 3 and 4 we present techniques for discovering invariants. These techniques were originally designed with an automatic implementation in mind. However, they are in fact also useful for finding invariants by humans. For simplicity of presentation, we consider the single block of Figure 2.3. We will distinguish between two general approaches to producing invariants:

(1) the *algorithmic approach,* in which we obtain guaranteed invariants $q(\bar{x}, \bar{y})$ at N directly from the assignments and tests of the loop (using also any already established invariants at L and at N), and

(2) the *heuristic approach,* in which we obtain a new candidate $q'(\bar{x}, \bar{y})$ for an invariant at N from already established invariants at N and old candidates which we have not yet been able to prove to be invariants.

3. Generation of Invariants: Algorithmic Approach

We present first the algorithmic approach for generating invariants. We distinguish between invariants derivable from the assignment statements and ones based primarily on the test statements. The input predicate $\phi(\bar{x})$ and the fact that a counter is always a non-negative integer will be used as "built-in" invariants whenever convenient.

3.1. Generating invariants from assignment statements

We observe that assignment statements which are on the same path through the loop must have been executed an identical number of times whenever the cutpoint is reached. Thus the counter n of the cutpoint can be used to relate the variables iterated. We denote by $y(n)$ the value of y the $(n + 1)$th time the cutpoint is reached since the most recent entrance to the block (assuming a local initialization of the counters). Thus $y(0)$ indicates the value of y the first time the cutpoint is reached.

We use a self-evident fact as the basis for generating invariants: for $\bar{x}$ such that $\phi(\bar{x})$ is true and for each path α around the loop, we have

$$R_\alpha(\bar{x}, \bar{y}(n - 1)) \supset \bar{y}(n) = r_\alpha(\bar{x}, \bar{y}(n - 1)) \quad \text{for } n \geqslant 1. \quad (2.1)$$

That is, if values $\bar{y}(n - 1)$ occurred at the cutpoint, and a path α around the loop is then followed [that is, $R_\alpha(\bar{x}, \bar{y}(n - 1))$ is true], then the next values of $\bar{y}$ at the cutpoint [i.e., $\bar{y}(n)$] will be the result of applying r_α to $\bar{y}(n-1)$.

In practice, if there is only a single path around the loop such as in the block of Figure 2.3, it usually suffices to ignore the path condition R_α, and find invariants which satisfy the stronger condition

$$\bar{y}(n) = r_\alpha(\bar{x},\bar{y}(n-1)) \quad \text{for } n \geqslant 1. \tag{2.2}$$

Considering (2.2) for each component of $\bar{y}$, we have a set of recurrence equations, one for each y_j. We now attempt to express as many as possible of these equations in *iterative form*, e.g., as

(a) $y_j(n) = y_j(n-1) + g_j(\bar{x}, \bar{y}(n-1))$ or
(b) $y_j(n) = y_j(n-1) \cdot g_j(\bar{x}, \bar{y}(n-1))$.

Such forms are desirable because they can be solved to obtain

(a') $y_j(n) = y_j(0) + \displaystyle\sum_{i=1}^{n} g_j(\bar{x}, \bar{y}(i-1))$ or

(b') $y_j(n) = y_j(0) \cdot \displaystyle\prod_{i=1}^{n} g_j(\bar{x}, \bar{y}(i-1))$.

There are two ways to obtain invariants at a cutpoint from equations of the form (a') or (b'). First, it may be possible to express

$$\sum_{i=1}^{n} g_j(\bar{x}, \bar{y}(i-1)) \quad \text{or} \quad \prod_{i=1}^{n} g_j(\bar{x}, \bar{y}(i-1))$$

as only a function of $\bar{x}$ and n, not containing any elements of $\bar{y}(i-1)$. We then have an assertion which relates $y_j(n)$, $y_j(0)$, $\bar{x}$, and n. Second, if for two variables y_l and y_k there is a relation between

$$\sum_{i=1}^{n} g_l(\bar{x}, \bar{y}(i-1)) \quad \text{and} \quad \sum_{i=1}^{n} g_k(\bar{x}, \bar{y}(i-1)),$$

or between

$$\prod_{i=1}^{n} g_l(\bar{x}, \bar{y}(i-1)) \quad \text{and} \quad \prod_{i=1}^{n} g_k(\bar{x}, \bar{y}(i-1)),$$